[英国] 戴维 · A.罗瑟里 著　刘成郗 译

卫星

牛津通识读本 ·

Moons

A Very Short Introduction

译林出版社

图书在版编目（CIP）数据

卫星 /（英）戴维 · A. 罗瑟里（David A. Rothery）著；刘成郗译.
—南京：译林出版社，2024.4
（牛津通识读本）
书名原文：Moons: A Very Short Introduction
ISBN 978-7-5753-0084-1

Ⅰ.①卫… Ⅱ.①戴… ②刘… Ⅲ.①卫星 Ⅳ.①P185

中国国家版本馆 CIP 数据核字（2024）第 048884 号

著作权合同登记号 图字：10-2018-429 号

卫星 ［英国］戴维 · A. 罗瑟里 / 著 刘成郗 / 译

责任编辑 陈 锐
特约编辑 茅心雨
装帧设计 景秋萍
校 对 戴小娥
责任印制 董 虎

原文出版 Oxford University Press, 2015
出版发行 译林出版社
地 址 南京市湖南路 1 号 A 楼
邮 箱 yilin@yilin.com
网 址 www.yilin.com
市场热线 025-86633278
排 版 南京展望文化发展有限公司
印 刷 江苏凤凰通达印刷有限公司
开 本 890 毫米 ×1260 毫米 1/32
印 张 9.875
插 页 4
版 次 2024 年 4 月第 1 版
印 次 2024 年 4 月第 1 次印刷
书 号 ISBN 978-7-5753-0084-1
定 价 39.00 元

序　言

戴　昱

　　当我们仰首凝望深邃的星空时，很难不被璀璨星辰与明亮行星的光辉所吸引。每一个天体都闪耀着自己独特的魅力：从大小麦哲伦如梦似幻的星云，到仙女座星系朦胧而神秘的边缘；从天狼星耀眼的光芒，到参宿四温暖的红光；从水星炽热的表面，到火星赤红的沙丘……然而，在这些天体巨人的阴影之下，我们将视线转向了更深邃、更细微的所在。那些散布于星系的万千恒星系统里，隐藏着无数鲜为人知的世界，那便是环绕行星运行的月球和卫星。它们与行星共同演绎着太空中的舞蹈，每一个都承载着属于自己的故事与待解的谜团，等待着我们去探索与发现。

　　阅读这本书，犹如启动了一场跨越时空的奇幻之旅。随着我们渐行渐远，地球的轮廓逐渐模糊，而卫星的多样性却愈发丰富。宇宙的辽阔与人类的微渺对比鲜明。然而，宇宙中从不缺少各种可能。自从系外行星首次被人类的目光捕捉，至今尚不足三十载。令人惊叹的是，在已探知的系外行星中，有近半数的恒星系统拥有不止一颗行星。这不禁引人遐想：在这无垠的宇宙中，是否存在着能够孕育出类地生命，甚至智慧生命的行星与卫星？或者，退一步讲，是否还有适宜人类定居与繁衍的卫星呢？

要回答这个问题，人类自然而然地将目光首先聚焦于离地球最近、夜空中最为璀璨夺目的天体——月球。自古以来，关于月球上居住着生命或神灵的传说在世界各地流传不息。“白兔捣药秋复春，嫦娥孤栖与谁邻。”广寒宫中月兔与嫦娥的故事或许是流传最广的版本。西方说法中“居住在月亮上的男子/女子/神祇”，他们所感受到的究竟是求长生被困的孤冷，还是守护地球的虔诚？

先进的望远镜与空间仪器比肉眼所见更清楚。它们所捕捉到的影像为我们揭示了月宫的真相——原来那里遍布着形态各异的环形山与陨击坑。随着数据的积累，月球上这些山海的物质构成、起源和演化历史也逐渐清晰。人类甚至已经多次亲自踏上月球表面，或者通过发送探测器的方式，从那里带回了数百千克的珍贵月壤。这些“纪念品”被带回地球研究和向公众展示，你或许就曾经在博物馆中一睹过月壤的真容。科学的认知让月球变得更加亲切，然而，围绕她的谜团依然众多，这令她依旧保留着一份神秘，始终令人向往。

月球是特别的，尽管开普勒创造的“卫星”（Satellite）一词如今已被广泛地使用，但即便是天文学家，也时不时会用“某某行星的月球”来指代其他行星的卫星。在我们的太阳系中，已发现的卫星数量已经远远超过了行星。除了地球，火星、木星、土星、天王星和海王星都拥有不止一颗卫星，甚至在冥王星等矮行星和越来越多的小行星周围，都发现了它们的卫星的身影。更令人惊讶的是，在有限的地外行星中，也发现了卫星的踪迹。

这一切的发现，主要归功于过去几十年来，光学和红外成像技术的飞速发展。从远距离观测的望远镜，到近距离探测的飞掠器，再到环绕轨道和直接登陆的探测器，探测手段和方法的日益成熟，使得我们获取的卫星图片也越来越清晰。在本书中，你将见到木卫一上的炙热火山、木卫二被冰层覆盖的巨大海洋、土卫二裂缝中喷射出的羽状物、天卫五上的崎岖悬崖……还有最晚被发现的火星的卫星——火卫一和火卫二，前者遍布着沟渠，后者则像一个大土豆。本书将带你深入探索这些天体的秘密，揭示它们引人入胜的物理性质、化学成分和复杂的地质构造，推演它们的形成和演化历史，这能让你更好地理解宇宙和生命的奥秘，感受那无尽的浩瀚与神秘。

除了深入探讨卫星的多种类型、有趣的命名方式以及太阳系卫星的性质与形成等，本书还致力于解答一系列引人好奇的问题。例如，“蓝月亮”是什么？月球的“背面”有什么？潮汐如何产生？每个月为什么是29天而不是22天？为什么有些卫星表面充满沟壑和陨击坑，有些却相对平滑？在零下200摄氏度的卫星表面上，会有怎样的奇特景象？此外，本书还致力于澄清一些广为流传的迷思与谣言。比如，“超级月亮”是否真的会影响人类的行为，或引发自然灾害？带着这些疑问，让我们一起踏上这场探索之旅吧。相信在阅读的过程中，你会收获满满的惊喜，对卫星及宇宙的奥秘有更深入的了解。

你最钟爱的卫星是哪一颗呢？作者分享了他心仪的几颗卫星。而对我来说，除了无可替代的月球之外，我的最爱是太阳系中唯一和地球一样，拥有大气层的土卫六。透过其厚厚的大气层迷雾，瞥见那隐藏其后、和地球如此相似的海洋、山谷、冰火山和湖泊等地貌时，我不禁会陷入想象：如果它的大气不是氮和甲烷，而是氮和氧的话，它是否就可以变成太阳系中另一个生机勃勃的地球呢？

不可否认，人类探索宇宙的深层动力，除了对未知世界的好奇心，必然也包含了一部分对寻找地外生命和人类未来家园的渴望。尽管目前观测到的卫星，绝大部分充斥着荒芜的地貌，但我们无法断言未来。随着人类的"视力"不断提高，在未来的某一天，说不定都不需要太久，我们就真的会在系外行星中意外地看到抱着玉杵的月兔了。

"今人不见古时月，今月曾经照古人。"李白千余年前发出的感慨，千年后的你我读来，仍能心潮澎湃，感慨万千。生命本身，在星体的变迁和宇宙的演化面前，实如沧海一粟。但思想的维度，却能超越宇宙的物理尺度，穿越时空的隔阂，达到亘古不变的共鸣。了解这些倏忽繁复中的事实和规律，或许就是我们迈向更高维度的第一步。

目 录

第一章

卫星的发现和意义

我们的卫星环绕着地球运行——从现在开始，我将称那颗卫星为“月球”，因为那就是它的名字。写关于我们如何发现“月球”的故事是毫无意义的。月球几乎和太阳一样明显。如果云层条件允许，我们几乎有一半的时间可以在傍晚的天空中看到它。如果我们醒得早，就能在其余大部分时间里看到它出现在黎明前的天空里。人们也经常能在白天发现它。

在最早有记载的时候，人类就已经很了解月球了，但在那之前肯定也一样，因为月球在夜间一定是一个受欢迎的照明光源。最古老的与月球相关的人工制品可能是刻有点或线的三万年前的骨板，一些人认为这是记录月相的一种方式，因为月球在29.5天的周期中从新月膨胀到满月，然后从满月收缩回新月。月球的外观之所以会发生这样的变化，是因为月球的运行轨道围绕着地球，它在天空中相对于太阳的位置不断地改变，所以我们可以看到被照亮的半球的面积也在变化。

要想确定是谁第一个意识到月球绕着地球转，也同样没有意义。对大多数古人来说，所有的天体都绕着地球转，这似乎是显而易见的。事实上这是错的，实际上，月球是天空中唯一环绕地球运动的自然天体。我这里指的不是我们看到的太阳和其他天体每天（在24小时中）升起和落下的运动——这只是由于地球在自转而表现出的“运动”——而是指月球相对于太阳的绕地球天空的运动，月球绕地球的天空运行一周为29.5天。这里的29.5天是由月球绕地球360度旋转一周需要的27.3天，再加上用于抵消地球在此期间绕太阳公转了十二分之一的两天多一点组成的。

随着时间的推移，围绕其他行星运行的卫星被发现，这表明就运动而言，地球并不是特别的。这为推翻16世纪至17世纪认为地球是万物之中心的根深蒂固的正统观念提供了关键证据。

虽然一些古希腊哲学家倾向于认为地球和行星围绕着一个“中心之火”（太阳）运行，但他们属于少数派。到17世纪初，经过长期确立的宇宙观把地球放在中间，而已知的天体围绕着它旋转。人们正确地认识到月球是其中距离最近的，然后是水星、金星、太阳、火星、木星和土星。再往外是一个上面承载着星星的天球。对大多数哲学家来说，这个天球也在旋转，因为地球似乎是静止的，尽管一些希腊人和印度人认为地球在自转。

希腊哲学家和他们后来的追随者都坚信天空应该是“完美的”，所以他们试图用均匀的圆周运动来明确地解释观察到的天体运动。随着观测精度的提高，在试图将理论与观测结果匹配时出现了越来越多的缺陷。这导致了复杂而烦琐的解释理论的发展。在其中的“本轮”系统中，较小的圆（本轮）被嵌套在较大的圆（均轮）上，并且只有在与相关圆轨道中心不重合的特殊点附近测量时，运动速度才是均匀的。

这种精心设计的以地球为中心的宇宙观（即地心说）通常被称为托勒玫体系，以一个名叫克洛狄斯·托勒玫的希腊化埃及人的名字命名，他于公元150年左右在亚历山大工作，并得到了天主教会的认可。在欧洲的大部分地区，宣扬与其相反的观点是危险的，而且地心说在中国和整个伊斯兰世界也占有主导地位。

然而，在16世纪早期，波兰天文学家尼古拉·哥白尼（1473—1543）提出了一个与之相反的理论，该理论认为，包括地球在内的行星是围绕太阳运行的（日心说模型），外围包裹着固定的恒星天球，只有月球绕着地球运行。这基本上是正确的（除了恒星不是固定在一个天球上——它们只是非常遥远），但为了使它符合现有的观测结果，哥白尼不得不引入比托勒玫体系所需的更多的本轮。直到约翰内斯·开普勒（1571—1630）在1609年引入椭圆作为围绕太阳运行的轨道，而不是正圆轨道，日心说模型才变得更加简洁。而到了1687年，艾萨克·牛顿（1643—1727）才用他的运动定律和万有引力理论解释了为什么轨道是椭圆的。

尽管哥白尼的模型为许多同人所知，但他不愿发表。直到他去世的1543年，他的巨著《天体运行论》才出版。与流行的说法相反，它并没有立即被教会禁止，但它的观点肯定是有争议的，因为它们与《圣经》中以地球为创世中心的宇宙观点相矛盾。

接下来是伽利略·伽利雷（1564—1642），一位在帕多瓦工作的意大利科学家。1610年，他将世界上最早的望远镜之一（由他自己建造）转向天空。他不但发现了金星的相位以及大量

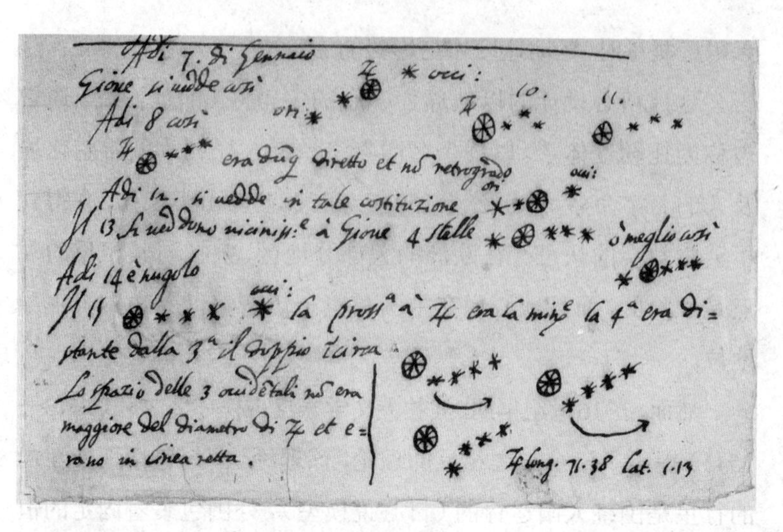

图1　伽利略笔记的一部分，连续记录了1610年1月7日至1月15日夜晚木星的4颗卫星（1月14日除外，那天多云）

暗淡的和肉眼看不见的银河中的星星，还看到伴随着木星的四颗暗淡的“星星”（图1），他也注意到它们在木星的两侧来回移动。仅经过了几个晚上的观察后，他得出结论：这些物体在一个距离他的视线非常近的平面上环绕着木星运动。

1610年3月，伽利略通过一本命名为《星际使者》的小册子宣传了他的观测结果。那年秋天，木星再次出现在天空中，包括英国的托马斯·哈洛特（1560—1621）和德国的西蒙·马吕斯（1573—1625）在内的几位天文学家证实了木星这四颗卫星的存在。事实上，马吕斯声称他自己在1609年就发现了它们。

在被称为“卫星”［其仍然是“月球”（moons）的正确学名］之前，这类天体的名称是“月球”。伽利略本人没有使用任何术语（至少一开始没有），仅把它们称为“星星”，尽管我们可以将这一称呼视为是在表示它们的点状外观，而不是表示他认为它们和普通恒星是一样的。马吕斯在他1614年的著作《木星》中称它们为“类木行星”，哈洛特也写了他自己看到“新行星”的经历。开普勒在1610年的时候首次把它们写作“卫星”。他用了一个拉丁语单词，意思是“陪伴更重要者的人”。

不管马吕斯的说法是否属实，伽利略都是第一个发表这一观点的人，因此他被认为是第一个通过观测证明所有天体运动并非都以地球为中心的人，而实际上其也并非都以太阳为中心。

卫星的利用

木星的卫星以一种可预测且有规律的方式围绕木星运行。然而，在发现木星卫星后的50年中，在巴黎工作的意大利人乔凡尼·卡西尼（1625—1712）注意到，卫星消失在木星阴影中的规律里存在着一些细微但系统性的差异。当地球的轨道靠近木星时，卫星消失（类似于日食）的时间间隔会稍微变短，而当两颗行星分开时，时间间隔会稍微变长。卡西尼正确地指出，这是因为光的传播速度并不像当时人们认为的那样是无限快的，实际上，光需要一个可测量的时间才能从木星到达地球。1676年，他提出，光穿过地球到太阳的距离大约需要10分钟到11分钟，这与正确的8分32秒非常接近。他的助手奥勒·罗默（1644—1710）很快就得出了观测结果，使更精确的估算成为可能，但这一基于卫星的光速有限的论证则直到大约50年后才被普遍接受。但在我们能够自信地将光速转换为熟悉的单位之前，还需要对地球与太阳的距离进行精确的测量。虽然卡西尼已在1672年估算过地球和太阳之间的距离，且只比1.5亿千米的正确值低7%左右。

在发现其他卫星之前，必须对望远镜进行改进，以超越伽利略和他同时代人建造的相当粗糙的设备。荷兰人克里斯蒂安·惠更斯（1629—1695）建造了一个更好的望远镜，并在1655年3月发现了土星的第一颗已知的卫星，将其称为“土星的月球”，这可能是第一次使用“月球”一词来指代另一颗行星的卫星。卡西尼在1671年到1686年之间还发现了另外四颗更暗（因此也更小）的土星卫星。

对当时的一些人来说，土星有五颗卫星，而木星只有四颗，这似乎是合乎逻辑的。然而，更多的发现接踵而至，它们破坏了这种一切遵循着简单模式的假象。1781年，在英国工作的汉诺威人威廉·赫歇尔（1738—1822）发现了天王星。1787年，赫歇尔发现了天王星的两颗卫星，1789年又发现了土星的两颗卫星。下一个被发现的卫星是海王星最大的卫星，是英国人威廉·拉塞尔（1799—1880）在1846年发现海王星后的第17天发现的。拉塞尔与美国的威廉·邦德（1789—1859）和乔治·邦德（1825—1865）父子共同被认为在1848年发现了土星的第八颗卫星，他还在1851年单独发现了天王星的另外两颗卫星。

火星仅有的两颗小卫星于1877年被美国的阿萨夫·霍尔（1829—1907）发现。1892年，美国人爱德华·爱默生·巴纳德（1857—1923）发现了木星的第五颗卫星，比伽利略发现前四颗卫星的时间晚了200余年。此处的延迟是因为这颗卫星远比其他卫星小，也更靠近木星，这两个因素都让它很难被看到。这是最后一颗被肉眼发现的卫星。随后的发现是通过摄影，第一颗的发现者是美国人威廉·皮克林（1858—1938），他于1899年通过前一年在秘鲁拍摄的照片找到了土星的第九颗卫星。最近的发现是通过安装在望远镜或宇宙飞船上的数码相机取得的。

所以，在我们的太阳系中有很多卫星——事实上比我刚提到的要多得多——但是了解它们有什么用吗？也许令人惊讶的是，伽利略发现的卫星甚至在17世纪就有了实际用途。在发明即使在运输中也能够保持准确时间的精密计时器（即时钟）之前（时钟发明于18世纪晚期），要确定经度是非常困难的。在陆地上，你可以尝试测量地点之间的距离，而在海上，你能做的最好的事情就是通过航向和速度来估算距离。这有时会酿成灾难性的错误，比如1707年，一队返航的英国海军舰队在锡利群岛附近触礁，造成1 400人死亡。

木星卫星的独立周期运动提供了一个部分解决方案，特别是在卡西尼制作了一套精确预测其卫星食的时间表之后。要确定经度，你所要做的就是测量一个特定的卫星食和当地正午之间的时间，然后用卡西尼的表格来确定你离卡西尼的参考经度以东或以西的距离。

卡西尼本人从法国国王路易十四那里得到了一项委托来改进其王国的地图，在17世纪70年代，卡西尼和他的团队开始确定法国主要城市相对于巴黎的经度。他们会观察木星的一颗卫星的卫星食，然后用一个摆钟（如果我们不去移动它，它就是一个完美的时钟）来测量从其发生到当地正午（当观察到太阳到达最高点时确定为正午）经过了多长时间。

有时候真相让人难以忍受。许多城市到巴黎的距离比预期的要近100千米，这表明法国从东到西的范围比之前所认为的要小。据说，路易曾抱怨，相比于他的敌人，他自己的天文学家夺走了他更多的领土。

卡西尼成功地将他的技术出口到海外。在航行的船上很难把望远镜对准木星，这使得它在海上实际操作起来很不现实，但它却可以用来在岸上测定经度，而且事实上在19世纪早期它还在被用作美国西部地图绘制的辅助工具。

月球本身也提供了另一种确定经度的方法，因为它每小时在天空中移动的距离大约等于它的直径。我们可以通过测量月球相对于一颗已知恒星的位置来确认时间，但我们必须校正视差，由于月球与地球的距离足够近，所以它相对于在天空背景中的恒星的位置从不同的地点看会略有不同。该方法的原理是由英国人埃德蒙·哈雷（1656—1742）提出的，哈雷后来因为在1683年左右出现的彗星而知名。必要的测量需要精准确定月球和恒星之间的夹角，然后进行复杂的计算，可能需要半个多小时才能完成；直到后来开发出一些表格，将这一过程缩短到大约10分钟。这种利用月球和参考恒星之间角差的方法被称为“月距法”，在1767年到1850年之间被广泛用于海上，那时人们已经可以买得起可靠的航海钟了。然而，在接下来的至少50年里，它继续作为一种后备技术被传授给航海家。

卫星还被用于测量它们围绕的物体的质量（或者更确切地说，是两个物体的质量之和）。这是因为轨道周期的平方正比于轨道半径的立方除以质量之和。如果卫星比行星小得多，那么它自己的质量是可以忽略的，所以轨道周期可以告诉我们这颗行星的质量。为了把这个测量值转换成熟悉的单位。我们需要将得到的质量除以一个被称为“引力常数G”的数字（G基本上告诉我们产生给定的重力需要多少质量）。

如果早一点知道G，那么科学家们就可以用月球的轨道周期来测量地球的质量。但事实并非如此，因为G直到1798年才由英国人亨利·卡文迪许（1731—1810）在一次实验中首次充分确定，该实验基本上同时确定了G和地球的质量。然而，一旦人们知道了G，太阳系的规模就确定了（与卡西尼的结果相差不大。由于100年后的1761年和1769年对金星凌日的观测，它的不确定性大大降低了），其他行星的卫星的轨道周期可以用来确定这些行星的质量和密度。例如，这揭示了木星和土星的巨大质量（分别是地球质量的318倍和95倍）以及它们的低密度（土星的密度只有水密度的69%）。

金星和水星没有卫星，因此它们的质量只能从它们对地球的轨道或恰巧经过的彗星造成的微小扰动来估算。这留下了很大的不确定性，直到航天器访问它们并近距离体验它们的引力。

卫星的命名

当人们开始发现卫星时，还没有给它们命名的系统。伽利略以其赞助人柯西莫·德·美第奇家族的名字将他发现的四颗卫星集体命名为“美第奇星”。他通过罗马数字Ⅰ、Ⅱ、Ⅲ和Ⅳ分别将它们区分开来，离木星越远数字越大。1614年，西蒙·马吕斯提议以伊俄、欧罗巴、伽倪墨得斯和卡利斯托，即宙斯（在希腊神话中等同于朱庇特）的情人的名字来命名它们。伽利略也许是对马吕斯在木星发现者——事上的敌对观点而感到愤愤不平，因而他对这些名字毫无兴趣。这些名字最终成为我们今天使用的官方认可的名称，尽管罗马数字系统仍在并行使用（也被用于最新发现的卫星）。然而，伽利略也很受尊敬，因为他发现的四颗卫星（比木星的任何其他卫星都要大得多）被称为“伽利略卫星”。

卡西尼把他发现的土星的四颗卫星命名为“路易星”，以他的赞助人法国路易十四的名字命名，接着是惠更斯发现的更大的“土星的月球”。其他的天文学家更喜欢使用伽利略风格的罗马数字来辨别单个的卫星。然而，从行星近处向远处编号的做法造成了混乱，因为新的发现可能会改变命名的顺序。例如，惠更斯发现的“土星的月球”先后使用了数字Ⅱ、Ⅳ和Ⅴ。在威廉·赫歇尔1789年的发现之后，国际社会一致认为，这个编号系统被冻结了，因此单个天体的名称再也不会被重新排序。1847年，威廉的儿子约翰·赫歇尔（1792—1871）为当时已知的土星的七颗卫星提出了一套统一的名称，我们沿用至今。他以克洛诺斯（土星之名萨图恩在希腊神话中的名字）的兄弟姐妹们的名字为其命名，其中最大的一颗被命名为泰坦，这是他们种族的统称，其他的以单独的泰坦巨人命名：伊阿佩托斯、瑞亚、特提斯、狄俄涅、恩克拉多斯和弥玛斯。当拉塞尔在1848年与其他人共同发现土星的下一颗卫星时，他将其命名为亥伯龙，即另一位泰坦巨人，这与约翰·赫歇尔的主题一致。

在他发现天王星的第三和第四颗卫星后，拉塞尔邀请约翰·赫歇尔将它们连同他父亲之前的两个发现一起命名。他选择了提泰妮娅和奥伯伦（他们得名于莎士比亚《仲夏夜之梦》中的仙后和仙王），爱丽儿（得名于亚历山大·蒲柏的《夺发记》中的天空精灵，她也出现在莎士比亚的《暴风雨》中），以及翁布里埃尔（《夺发记》中的忧郁精灵）。

令人惊讶的是，拉塞尔并没有命名他在1846年发现的海王星卫星。它现在的名字，特里同（海王星之名涅普顿在希腊神话中是波塞冬，特里同为波塞冬之子），直到1880年才有人提出。

火星的两颗小卫星，火卫一和火卫二的名字是由它们的发现者根据伊顿公学的科学老师亨利·马丹（1838—1901）的建议而选择的，他的建议来自于《伊利亚特》第十五卷，在这本书中，阿瑞斯（在希腊神话中等同于马尔斯）召唤了孪生兄弟福波斯（意为恐惧）和得摩斯（意为恐怖）。

后来发现的卫星（火星已经没有更多卫星了）的命名，遵循了在每个行星的首个卫星的名字被普遍接受时确立的主题。国际天文学联合会（下称IAU）自1919年成立以来，一直是太阳系天体名称（及其拼写）和表面特征的仲裁者。一颗新发现的卫星会被授予临时的称号（例如，S/2011 J2是2011年发现的第二颗木星卫星），只有当它的轨道能被很好地描述时，才能得到正式的名称。木星卫星的名字取自宙斯（朱庇特）的恋人和后代。幸运的是，神话提供了大量的名字可供选择，因为木星有50颗已命名的卫星，还有16颗仍在等待正式命名。

土星有相似数量的已知卫星，其中53颗已经被命名。最初的18颗以泰坦巨人和他们的后代命名。约翰·赫歇尔最初定下的主题后来扩大到包括希腊、罗马、高卢、因纽特和挪威神话中的其他巨人，根据它们的轨道进行分配。

除了贝琳达是另一个来自蒲柏的《夺发记》中的名字之外，天王星的卫星持续从莎士比亚笔下的角色名（大部分是女性）中获取名字，从异国情调的希克拉库斯（《暴风雨》中卡利班的母亲）到平淡无奇的玛格丽特（《无事生非》中的女仆）。海王星的卫星被分配了与波塞冬（涅普顿）有关的神话人物的名字。

有多少卫星？

我们很难追踪太阳系中已知的卫星数量。自2000年以来，我们已经有了大量的发现。这些卫星主要是由地面望远镜和哈勃太空望远镜发现的，但美国航空航天局（下文简称NASA）的“卡西尼号”航天器在土星轨道上发现了土星的几颗小卫星。已知拥有卫星的天体包括从地球到海王星的所有行星。似乎可以肯定的是，金星和水星这两颗内行星都不可能有一颗直径超过一千米左右的卫星，否则它就会被望远镜或轨道航天器发现。

我们太阳系已知的190颗行星的卫星列在附录的表格中。这里我引用“平均半径”主要是因为较小的卫星，特别是那些半径小于200千米的卫星，在形状上可能明显不是球形的。相比之下，地球的平均半径是6 371千米。我列出的密度单位是千公斤每立方米，这相当于吨每立方米，或者说克每立方厘米。为了比较不同的卫星，在使用相同的单位时，水的密度是1.0。

海王星之外的各种冰天体也有卫星。其中，冥王星有已知最大的卫星群，具体可见附录中的表格。类冥王星的天体体积庞大，其中一些天体有卫星也就不足为奇了。然而，一些小行星，包括一颗直径不到1千米的小行星，也被发现有卫星（根据定义，卫星甚至比它们所围绕的物体还要小）。很少有人预料到这一点，我将在第七章中再次讨论这个话题。

地球有其他卫星吗？

你可能会在网络上或考试中发现这样的说法：地球除了月球还有一两个小卫星。这些都是错误的。虽然太空中有很多人造卫星，但在围绕地球的永久轨道上还没有发现小型的自然物体。它们的直径必须小于10米，否则早就被发现了，而且计算表明，这样的“迷你卫星”不可能有稳定的轨道。

然而，也有一些小行星，它们的轨道与地球的轨道相交，这使得它们能够暂时被捕获，在此期间它们确实是小卫星。一个例子是5米宽的“2006 RH_{120}”天体。这是2006年用望远镜发现的，接下来的一年，它绕地球转了四圈，每一圈的形状都不同。它的运行路径几乎远远超出了月球的轨道，但它最接近地球的距离达到了月球与地球距离的70%左右。作为地球的临时卫星，它在11个月后脱离了地球。它的轨道一直很复杂，不稳定，部分原因是它会同时受到月球和地球的引力。到2017年，它将在太阳的远端，但将在2028年再次接近，为临时被地球捕获提供了另一个机会。接下来会发生什么是无法预测的，因为像这样的不规则形状的物体足够小，足以让太阳辐射的光压影响它们的轨迹。我们无法估计辐射造成的扰动大小，因为我们既不知道物体的形状也不知道它的密度。

实际上，“2006 RH_{120}”可能不是一个自然物体；它可能是美国宇航局1969—1972年登月计划中使用的阿波罗火箭之一的第三段。尽管如此，一些研究表明，在任何时候，地球都可能会伴随着多达几十个临时被捕获的小于两米大小的轨道天体，它们到达，环绕至少一个周期，然后消失。更大的例子比较罕见，可能每10万年才会有一次100米大小的物体以这种方式被暂时捕获。

另一类有时被误报为地球的“卫星”的天体是小行星，它们围绕太阳公转的平均轨道周期与地球的公转轨道周期完全相同。这样一颗小行星围绕太阳的轨道非常靠近地球的轨道，但地球的引力对它产生了强烈的影响，因此它以如下的重复模式循环：想象一下，这颗小行星沿着比地球轨道稍微靠近太阳的轨道运行。这意味着这颗小行星绕太阳运行的速度将比地球略快，最终将会赶上地球。当它靠近时，地球的引力将小行星拉入一个更大的轨道，更大的轨道需要更长的周期，所以小行星现在围绕太阳运行的速度更慢，最后落在地球后面。最终，在它们都绕轨道运行了几圈之后，地球几乎追上了小行星，又把它拉进了一个更小的轨道。现在这颗小行星以比地球还快的速度向前移动，又回到了最开始的模式。

从地球的角度看，这颗小行星的轨迹类似于一个马蹄铁（地球在马蹄的开口部分）。正因为如此，这些轨道有时被称为“马蹄形轨道”，但重要的是要认识到，小行星并不是围绕地球运行，而是围绕太阳向前运行。至少有三颗这样的小行星是已知的，其中最大的“2010 SO_{16}”的平均半径约为300米。其他小行星的轨道不像地球那么圆，周期也与地球相似，这导致它们在地球前面或后面的平均位置移动，而不是完全绕着一个马蹄形移动。其中最著名的是“3753克鲁斯娜”，它的平均半径约为2.5千米。它也是绕太阳运行，而不是地球，更不是月球。你可以在拓展阅读的链接中找到更多关于这种“准卫星”的奇怪轨道的信息。

卫星能拥有卫星吗？

行星绕着太阳转，卫星绕着它们的行星转，所以一个很自然的问题是，任何一个卫星是否可以有自己的天然卫星——卫星的卫星。就我们的月球而言，近年来，人类在月球轨道上放置了几颗人造卫星作为临时卫星。然而，它们都处于不稳定的轨道，这将导致它们在几年后撞击到月球表面。这是因为月球的引力不足以控制它周围的空间区域，因为质量大得多的地球离它太近了。

月球围绕地球的轨道是稳定的，因为地球的引力足够强大，足以在小于约100万千米的距离内胜过太阳的引力，这个空间体积被称为地球的“希尔球”，以美国天文学家乔治·威廉·希尔（1838—1914）的名字命名，因为他定义了这个概念。月球的轨道就在这里面，因此它具有长期的稳定性。月球自身的希尔 15 球直径为6万千米，但任何围绕月球运行的物体，即使是在它的希尔球内，也会受到地球的足够引力，导致其轨道随着时间的推移而缩小。事实证明，其他天体的卫星也是如此，所以太阳系任何地方的卫星轨道都不是长期稳定的。根据一种复杂的力的平衡，围绕月球运行的轨道的持续时间从几年到数百万年不等，但比太阳系45亿年的年龄要短得多。

因此，不存在“卫星的卫星”，如果发现了一颗，这一状况几乎肯定是短暂的。 16

第二章

月 球

现在我们来详细地看看一些单独的卫星。我将从月球开始，因为这是最广为人知的一颗，当然，你可以亲眼看到它。

月球的英文名“moon”最早可追溯到日耳曼语（在古英语里为Mōna，原始日耳曼语中为Mǣnōn）。在拉丁语中，它被称为“Luna”（即英文中的形容词lunar，法语中的Lane，西班牙/意大利语的Lana，以及发音相同的俄语Луна的词源）。在古希腊，月球是Σελήνη（Selene），由此我们得到了“selenographic coordinates”（月面坐标，用于绘制月球的经纬度系统）等术语的前缀“selen”。“moon”这个词最终被类比为指代任何围绕另一颗行星运行的物体。尽管正如我们所见，用开普勒创造的“satellite”（卫星）这个词来指代它们要便捷得多。

月球是一个巨大的天体。只有四颗卫星的体积比它大（同样质量也比它大）：木星的木卫一、木卫三、木卫四和土星的土卫六。如果月球能独立地绕着太阳公转，毫无疑问，它将跻身于“类地行星”之列，类地行星是最靠近太阳的四颗行星。表1显

17

示了这些天体的赤道半径和极半径，因为它们的旋转扭曲了它们的形状，所以它们在赤道处隆起，在两极处扁平。这使赤道半径大于极半径。金星是个例外：它的自转速度比地球慢近250倍，所以没有显示出可测量的扁平化。

表1　月球和类地行星的比较

名称	赤道半径（千米）	极半径（千米）	质量（10^{24}千克）	密度（10^3千克/立方米）
月球	1 738.1	1 736.0	0.073 42	3.344
地球	6 378.1	6 356.8	5.972 6	5.514
水星	2 439.7	2 437.2	0.330 1	5.427
金星	6 051.8	6 051.8	4.867 6	5.243
火星	3 396.2	3 376.2	0.641 7	3.933

尽管月球不比水星小多少，但由于其总密度较低，其质量要小得多。这是因为水星有一个非常大且密度很高的富含铁的地核，周围环绕着相对较薄的岩石地幔和地壳，而月球几乎完全是由岩石构成的。它的地核，如果有的话，半径也不超过300千米。

水星在其地核的外围有一个厚厚的熔融区。它是在不断循环的，又由于它是一个导体，所以就像发电机一样，产生了一个环绕星球的磁场，部分保护了星球表面免受来自太阳的带电粒子（宇宙射线）的轰击。月球缺乏这样的磁场，它的表面暴露在太阳投射出的任何东西下。

和水星一样，月球自身的引力太小，无法保留形成大气层所必需的气体，但月球表面仍有一些气态原子。例如，当表面被太阳风击中时，钠和钾原子通过“溅射”释放出来，氦是由太阳风直接带来的，而氩则是由钾的同位素的放射性衰变产生，并从月球内部逃逸出来的。这些原子的总“大气压”大约是地球表面大气压的3×10^{-15}倍（三百亿分之一！），这一数值如此之低，以至于原子更有可能逃逸到太空中，而不是相互碰撞，这使得月球的整个大气层相当于地球大气层中非常稀薄的最外层区域，即所谓的“外逸层”。

由于几乎没有大气层，月球表面的昼夜温差很大，中午赤道附近的温度为120摄氏度，晚上则为零下150摄氏度。这些昼夜变化不会深入月球土壤，即“风化层”中。在月球地表以下约1米的地方，温度被认为是相当稳定的零下35摄氏度。在两极附近有一些环形山，它们的底部从来没有被太阳照亮过，故其表面温度永远低于约零下170摄氏度。

月相，轨道和自转

绝大多数人都非常熟悉月球的外观。即使用肉眼，我们也能看到它表面的黑斑。第一批望远镜使伽利略和哈洛特（他做得更好）等观测者能够绘制出这些暗斑，也能辨认出更小的特征，如陨击坑和山脉。如果你用一副双筒望远镜来看，你可能会比他们看到更多。

无论何时，你都总是能看到与哈洛特和伽利略所绘制的相同的月球半球，因为月球的同一面总是对着地球。而不同的是，我们可以在任意特定的时间看到它被太阳照亮了多少，这取决于月球在其29.5天的轨道周期上的位置。当月球位于地球和太阳之间时（我们称之为“新月”），我们根本看不到它。它在极少数情况下会在地球和太阳正中间出现，从而导致日食，因为月球绕地球的轨道相对于地球绕太阳的轨道倾斜了约5度，因此月球经过太阳时通常不是略高于太阳，就是略低于太阳（我们都看不见），而不是正对着太阳。

新月过后的几天，在地球的天空中，由于月球与太阳的距离已经足够远，所以可以看到它变成一个细长的月牙，月牙不断充盈，直到面向地球的半球的一半被完全照亮。令人困惑的是，它被称为“上弦月”，指的是经过了29.5天周期的四分之一，而不是月球的圆盘被照亮了四分之一。可见的被照亮的区域继续增长（它的形状现在被称为“凸月”），直到新月后近15天，月球到达离太阳较远的那一边。这时，面向地球的半球被完全照亮，我们称之为“满月”（此时如果地球正好挡住了它，使阳光无法直接照射到月球，就会出现月食）。随着月球继续在其轨道上运行，被照亮的部分逐渐缩小，直到只有面向地球的半球的一半被照亮（即“下弦月”），然后它变为一个逐渐亏损的月牙，直到接近下一个新月时消失。

如果你认为你一直看到同一面就意味着月球不会自转，那你就错了。为了保持月球的同一面始终对着地球，月球必须每个月自转一圈。我在拓展阅读中放了一个链接，里面有一个动画对它进行了演示。几乎所有已知的卫星都处于这种“同步自转”（即潮汐锁定）的状态，因为潮汐阻力迫使它的自转与轨道公转同步。

这是因为两个相邻天体之间的相互引力，比如行星和它的卫星，会略微扭曲它们的形状。在每个面向对方的半球的中部会发生潮汐隆起（在月球上大约为100米）。这是因为靠近对方的一侧比其中心更接近另一个天体，因此受到的引力稍许更强。

20

在远离对方的另一侧有一个同样的凸起，因为天体的中心受到的引力比它的远端更加强烈。

如果月球的轴向自转周期比它的轨道周期短，则潮汐膨胀会在全球范围内变动，月球的形状将会不断扭曲，这样才能与行星保持一致。这将不断消耗能量，直到月球自转放缓到与轨道周期相一致。因此，尽管月球和其他行星的大多数卫星一开始可能自转得更快，但它们的自转现在与轨道周期相匹配。

顺便说一句，尽管月球有近地面和远地面，但它并没有永久的暗面。“月亮的暗面”只是一个为人们所接受的比喻，指的是我们曾经一无所知的隐藏的月球的那面。然而，由于月球的自转，在单一轨道的运行过程中每一面都将会面向太阳——当然，太阳在任何时候都只能照亮月球的一半。

虽然月球绕着地球公转，但它同时也在一年的周期中伴随着地球绕太阳公转。地球绕太阳公转的速度约为每秒30千米，比月球绕地球公转的速度快得多：月球最近时（即近地点）的公转速度为每秒1.022千米，最远时（即远地点）的公转速度为每秒1.076千米。因此，月球绕太阳运行的轨道分布在地球轨道的两侧，但它接近太阳的一面总是更凸出。它永远不会回到离太阳更远的地方。

月　面

我们是第一代知道月球背面是什么样子的人。直到1959年10月，苏联的“月球3号”探测器被发射到月球背面，传回了第一批模糊的照片，它才为人所知。图2从四个不同的方向展示了月球，以更清晰的现代视角显示出月球背面。近地侧视图显示

图2　由NASA的月球勘测轨道飞行器拍摄的四张月球照片组合而成。近地侧（以0度经度为中心），远地侧（以180度经度为中心）和两个中间视图（东侧，以90度经度为中心，西侧，以270度经度为中心）

了我们在满月时从地球上可以看到的月亮。远地侧视图显示的是月球的背面完全被太阳照亮时的样貌。

近地侧和远地侧之间有明显的差异。大约有一半的近地面被黑色的斑块所占据。在17世纪，天文学家把它们错当成海洋（或者至少是昔日海洋干涸后的海底），并用拉丁语中的海洋

22 “mare”表示它们，其复数形式为“maria”，即月海。我们现在知道，这些地区实际上曾被由类似玄武岩的大量火山熔岩所淹没，那里从来没有存在过水。但“月海”这个词今天仍在被使用，既指它们的一般名称，也指它们的正式名称（得到了IAU的批准）。“mare”和“maria”的正确读音是“MAH-ray”（而不是雌马的读音“玛雅”）和“MAH-ria”（而不是女孩名字的读音“玛丽亚”）。

在月球背面，只有几个相对较小的区域被月海所占据。据信，一部分原因是月球背面的地壳更厚，因此岩浆更难到达地表。它还表明，至少从38亿年前，当月海中大部分开始充满熔岩时，月球就一直在以同一个半球面向地球的状态同步自转。

在月海之外，月球表面较亮的部分是一种不同的岩石。它主要由一种叫作斜长石的长石矿物组成，因此这种岩石被称为“斜长岩”。它代表了月球最古老的地壳，有45亿年的历史。这种地形被称为“月球高地”（又称“月陆”）。近地侧月陆的低洼地区（大多数是受陨石撞击导致的大盆地）被月海的玄武岩淹没，但这种情况在远地侧没有发生。

从地球上，我们可以轮流看到月球远地侧的两侧边缘，因为月球的轨道不是正圆形的，而是椭圆形的，这使得月球中心离地球中心的近地点距离只有36.33万千米，而在远地点距离是40.55万千米。轨道速度越近越快，越远越慢，但月球的自转速率保持不变。因此，在远地点附近的自转速度超过了轨道运动速度，所以我们可以看到月球平均前端周围少许距离内的地形，而接近平均后端的地形则由于月球自转从视野中消失。当接近

23 近地点时，我们可以看到稍微超出平均后端的地形。总的来说，

从地球上能看到约59%的月球表面，尽管对于靠近边缘的地形我们只能看到一个压缩的、高度倾斜的图像。

这种月球外表左右摆动的现象被称为“天平动”，如果你点击拓展阅读的另一个链接，你就可以看到它的动画。它还显示了月球在近地点看起来是如何稍微更大一些（因为它离得更近），以及从地球上看，月球被照亮的那一半的可见形状（即所谓的“月相”）是如何随着月球绕其轨道运行而变化的。

图2中的图像没有任何阴影，因为拍摄时月亮在几何上正对着太阳。这就放大了不同表面反射的阳光比例的差别（一种被称为“反照率”的特性），但地形则并不清晰。图3展示了一个凸月的视图，其中越靠近明暗分界线，也就是区分月球上的黑夜和白天的分界线，阴影就越明显。 24

特别的是，由于其边缘的阴影，在明暗分界线附近可以看到许多圆形坑。这些都是环形山，我将在下文中讨论它们的起源，但首先看看它们在图3和图2的近地侧视图之间的外观有什么不同。图3左上方最明显的环形山出现在黑暗的背景上，因为它位于月海区域。这是一个直径达93千米的环形山，被命名为“哥白尼环形山”，它被认为是在大约8亿年到11亿年前形成的。它在图2的近地侧视图中非常突出，因为它的底部穿透了低反照率的月海玄武岩，进入了深埋的高反照率的月陆地壳。

图3左下角的一个显眼的环形山位于图2中一个亮斑的中间。这是一个直径达86千米的环形山，名叫第谷。它有一个比哥白尼环形山更明亮（更高的反照率）的底面，因为那里没有玄武岩，都是新的粉末状月陆地壳。一些几乎同样明亮的喷出物覆盖了环形山周围的表面，在这之外有明亮的射线向外辐射，这

图3　从地球上看，此时月球大约离满月还有3天8小时。明暗分界线在左边，而右边可以看到月球被照亮的边缘，即与月球背面的交接处

是细粉状喷出物的条纹，称为射纹。“阿波罗17号”在第谷环形山东北约2 000千米处着陆，但宇航员收集了穿过着陆点的一条第谷射纹的碎片。在地球上进行的分析发现，这些粒子大约有1亿年的历史，这表明第谷环形山比哥白尼环形山要年轻得多。

在图2的四张图片中，可以看到其他几个有射纹的环形山。只有当太阳接近正上方时，这些射纹才清晰可见。它们会随着

时间的增加而褪色，所以只有最年轻的环形山才有。哥白尼环形山也有射纹，但没有第谷环形山的那么明显。 25

陨击坑

陨击坑是布满月球表面的圆形边缘的洼地。事实上，它们的数量在太阳系大多数固体天体的表面都很丰富，包括在几乎每一颗卫星上。只有在古代的地表特征被侵蚀抹去，或者被熔岩流或风吹起的尘埃掩埋的地方，它才比较少见。

以地球为例，地壳在板块构造的作用下循环，板块构造在碰撞中挤压大陆的边缘，将古老的海底推入地幔。这增加了侵蚀和掩埋作用产生的影响，所以地球上很少有明显的像月球上那样的陨击坑。虽然现在已经有近两百个撞击坑被记录下来，但很少有非常壮观的，绝大多数都被侵蚀了，当人们开始注意和推测月球陨击坑时，没有一个地球陨击坑为人熟知。

是伽利略第一次称它们为“craters”（环形山）的。这是一个希腊单词，意思是饮用容器。它们在月球上的存在反驳了月球作为一个天体应该有一个完美光滑表面的观念——即使这个观念在那时肯定也已经很难证明了，因为即使没有望远镜，也可以看到月球存在明显的暗斑，显示出月球存在“污点”，而我们现在知道那是月海。

几个世纪以来，月球陨击坑的起源一直是一个有争议的问题。大多数科学家认为它们是某种火山现象，比如由火山爆炸性喷发造成的洞，或者只是巨大的气泡破裂留下的疤痕。最常见的替代理论是，它们是外部物体撞击月球留下的疤痕，但这一理论在大多数时候并不流行。这一理论的批评者指出，除了极 26

少数例外，月球的环形山的轮廓都是圆形的，而撞击物应该以各种角度撞击月球，因此我们可以预计陨击坑有很大一部分会呈不对称的细长形。另一些人反驳说，月球上的环形山与我们看到的大多数陆地火山非常不同，所以火山起源也不符合观测结果。

以现在的视角来看，有着无数的岩石碎片正以高速穿越太阳系，这似乎是显而易见的。但直到1808年，美国总统托马斯·杰斐逊（绝不像他的一些继任者那样对科学一无所知）还公开对此表示怀疑。在耶鲁大学的代表对一颗在康涅狄格州被发现坠落并被收集起来的陨星的调查结果中，他曾宣称他“宁愿相信两个美国教授会撒谎，也不愿相信石头会从天堂掉下来”。

20世纪60年代，对月球陨击坑的撞击假说得到了证据的坚定支持。有三件事有助于解决这个问题：第一，美国人尤金·吉恩·舒梅克（1928—1997）研究了亚利桑那州直径1.2千米的巴林杰陨星坑（也被称为流星陨击坑），发现矿物石英已经转化为只有在高压下才能形成的密度更大的二氧化硅。他在地下核试验场发现了相同的矿物变化，并正确地得出结论：巴林杰陨星坑是由一颗飞行物体以每秒数十千米（这是预期的轨道相交时相对速度）撞击地面时产生的高压冲击波所造成的。第二，由无人探测器传回的月球表面特写图像显示，月球表面有小到可见的最小尺寸的环形山。这与任何一种火山现象都不一致，而与物体撞击一个真空中的物体却不会有任何区别。第三，实验室实验的结果是，当一个“超高速”弹丸在真空室中射入目标时，27形成了看起来令人信服的圆形轮廓的弹坑。

图4是三个一系列不断嵌套且分辨率越来越高（即越来越详细）的月球陨击坑的视图，从哥白尼环形山自身开始。值得注

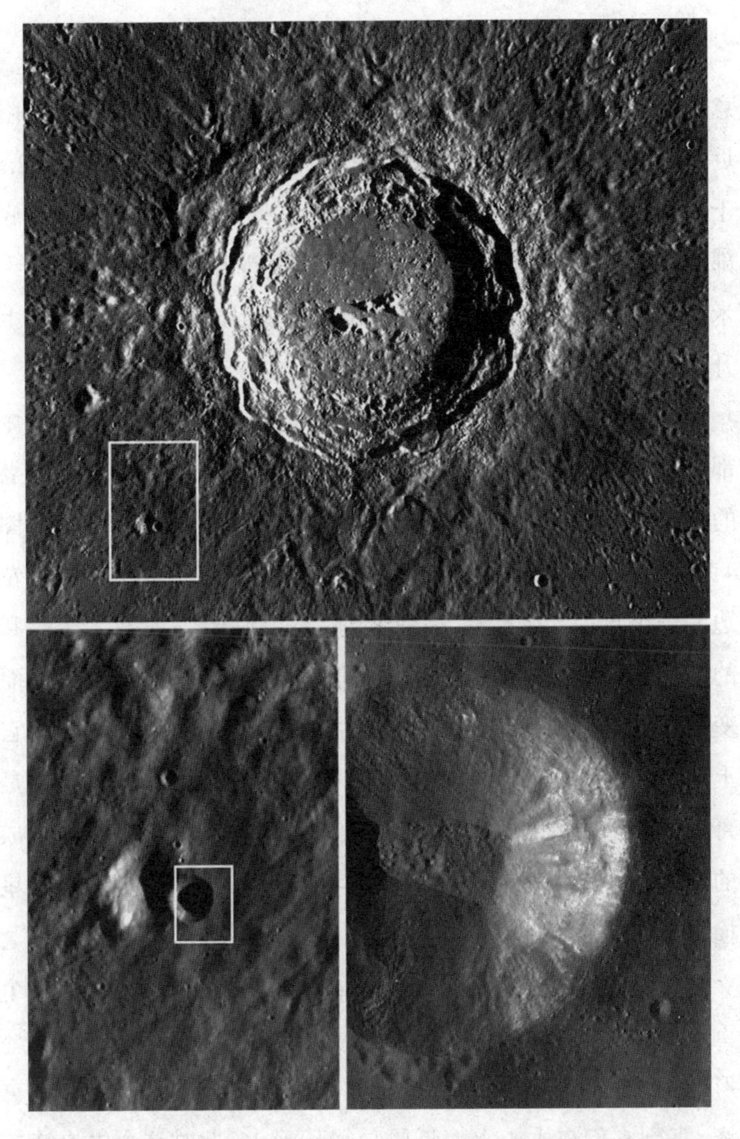

图4　哥白尼环形山与月球勘测轨道飞行器以不同的分辨率看到的周边区域。最上面是哥白尼环形山本身，比例尺是每像素250米，所以这幅图像的直径约为200千米，而太阳在东方较低的位置。左下角的高分辨率视图显示上方框内区域。左下为20千米宽的区域，比例尺为每像素64米。右侧的高分辨率视图显示左下方框内区域，同时7千米宽的哥白尼B环形山就在这个方框的西边。右下方是比例尺为每像素8米的未命名的3.3千米宽的环形山，拍摄时太阳在更西方天空中更高的位置

意的特点是，哥白尼环形山的坑底是平坦的，除了在它的中心附近有一组山峰。环形山的底部比环形山周围平原的高度低2.5千米，而边缘是高出平原1千米到1.5千米的凸起，故从边缘到底部的最大落差约为4千米。中央的山峰高出环形山底部约1千米，所以它们的峰顶一直在环形山边缘以下，实际上甚至没有上升到周围平原的水平。

在哥白尼环形山之外，一个环形山半径以上距离内的区域都被环形山形成时抛出的喷出物覆盖着。这掩埋了所有古老的环形山，但你看得越深入，你能看到的小环形山就越多。图4的中等分辨率视图（即左下图）的中间就有两个相邻的。左边的一个直径约7千米，被命名为“哥白尼B”。它有一个柔和的形状，因为它的轮廓被埋藏在哥白尼环形山的喷出物下面而变得平滑。另一个未命名的直径3.5千米的环形山重叠在它上面，并且有一个更清晰的外观。这个环形山更年轻，要么是哥白尼环形山的一大块喷出物击中地面时的“二次撞击”形成的，要么是在后来的某个时间点，由一颗不相关的小型小行星撞击形成的。

哥白尼B和它的邻居在其中心都没有山峰，这说明了一个重要的观察结果：在月球上，小于10千米至20千米大小的环形山中没有中心山峰。比这更大的环形山通常只有一个中央峰，直到它们在大小上接近哥白尼环形山，此时这个中央峰被一群山峰取代。在更大的尺寸下——直径超过350千米的环形山——其结构就变成了双坑甚至是多个坑。

图4中最高分辨率的视图是用每像素8米的比例尺记录下来的，包含了哥白尼B环形山所重叠的大部分较年轻的环形山，

大约为300米深。这张照片拍摄时太阳在西边，且比其他两张照片拍摄的要高得多，所以环形山的底部没有阴影。在这样的细节层面上，你可以看到环形山内部的斜坡缺少在哥白尼环形山中所看到的那种阶梯结构。在环形山外，你可以辨认出小到直径约20米的岩石和直径约50米的环形山。

毫无疑问，月球上绝大多数的环形山都是由撞击造成的。小行星是岩石或富含铁的天体，通常以每秒17千米的速度撞击月球。而彗星主要是来自外太阳系的冰构成的天体，它们的撞击速度往往更快，约为每秒50千米。

一旦撞击物以这样的速度撞击月球表面，冲击波就会从撞击点放射出来，这就是为什么环形山是圆形的，而且撞击角度几乎不重要。冲击波将撞击目标的物质和撞击物本身熔化并破碎，并将破碎和熔化的物质作为喷出物向外抛去。最先被抛出的喷出物速度最快，抛出距离也最远。在冲击波“挖掘”地面的过程中，环形山变得更宽而不是更深。随着撞击能量的消耗，最后的喷出物几乎无法飞出边缘，从而造成了边缘的抬升。相比于更早被抛出的喷出物，较晚出现并落在环形山附近的喷出物来自更深层的地方，所以撞击点物质的任何分层在沉积的喷出物中都是相反的。

在足够大的环形山中，中心山峰通过一种名为“回弹隆起”的过程形成。在高速拍摄的水滴撞击池塘表面的视频中，你可以看到类似的回弹，而不同的是在撞击形成的环形山中，山峰在有机会下沉之前就固定了。

环形山最终会比造成它们的撞击物宽30倍到100倍，而且整个过程非常快。在典型的撞击速度下，要像哥白尼环形山那样形成一个直径为93千米的环形山，需要一颗直径约为5千米的岩石小行星或一颗直径约为4千米的冰彗星，且大约需要几分钟完成。像这样大的环形山的内壁并不稳定，随后会坍塌成一系列同心阶地。要像哥白尼B环形山那样形成一个直径约7千米的环形山，只需要直径200米的小行星或直径100米的彗星，而且在大约30秒内就能完成。

对我们来说幸运的是，因为它们撞击地球的频率比撞击月球的频率还要高，所以能够制造出哥白尼环形山大小的环形山的撞击物现在已经非常罕见了。然而，在太阳系历史的最初10亿年里，地球周围有更多的碎片，其中一些大到足以形成数千千米宽的环形山（通常被称为陨击盆地）。地球上已经没有这些东西的痕迹了，但有许多在月球上保留了下来。

在月球的近地侧，这些巨大的环形山随后被熔岩淹没。圆形的轮廓证明了它们的起源，尽管近地侧的一些盆地已经被熔岩完全填满，导致被淹没的地区已经和周围地区融合。在图3和图2左上方的近地侧视图中，你应该能够辨认出直径为1 146千米的雨海。沿着它的东南边缘有一条弧形的山脉，这是盆地边缘的遗迹。如果你在图2中看西侧视图，你会看到赤道以南的东海盆地。它有一个直径约为500千米的内环，部分由月海玄武岩填满，而其直径为920千米的外环大部分是不含玄武岩的。近地侧的月海很可能都是双环，或者像这样的多环结构，但它们已经被淹没得如此之深，以至于在表面上看不到内环的痕迹。

东海盆地位于近地侧和远地侧的边界上。完全在远地侧中的最大的月海是莫斯科海，在图2中远地侧视图的左上方，它填满了一个直径近为500千米的双环盆地的内环，并溢出到外环的一部分。在远地侧还有一个大得多的盆地，其中只有少量或根本没有月海玄武岩。这是直径为2 500千米的南极-艾特肯盆地，是月球上保存下来的最大、最古老的盆地。它的深度为13千米，在图2远地侧视图中显示为一个中等反照率的区域（比月陆暗，但比月海亮），从南极向北延伸。

环形山与时间测定

月球上的环形山对于研究影响月球的一系列事件非常有用。解决这个问题只需要常识。例如，一个叠加在另一个环形山上的环形山一定是较年轻的，而且（更有用的是）一个环形山散布的喷出物和另一个被喷出物覆盖的环形山之间的相互关系可以显示出，这两个环形山中哪一个是年轻的。环形山和月海玄武岩之间也存在可用于判断的关系，如图5所示。

在这张图片中可以看出很多东西。现在，我将只讨论这些环形山和充满整个视野的月海玄武岩之间的关系。贝克托夫环形山和那些标记为D、E和L的环形山是简单的碗状环形山，毫无疑问，它们比月海玄武岩的表面要年轻。然而，扬松环形山呢？它的直径是24千米，大到足以形成一个中心山峰。然而，它的平坦的底面只比边缘低150米。对此的解释是，当月海玄武岩淹没这一地区时，扬松环形山已经在那里了。玄武岩熔岩流漫过环形山边缘，淹没并掩埋了环形山中央的山峰。随后的脱气和热收缩导致熔岩表面下沉，这使得被掩埋的环形山边缘（但不是深埋的中央山峰）在地形上重新显现出来。Y环形山在扬松环形山充满熔岩的底部又加深了550米，是由很久以后的一次撞击造成的。

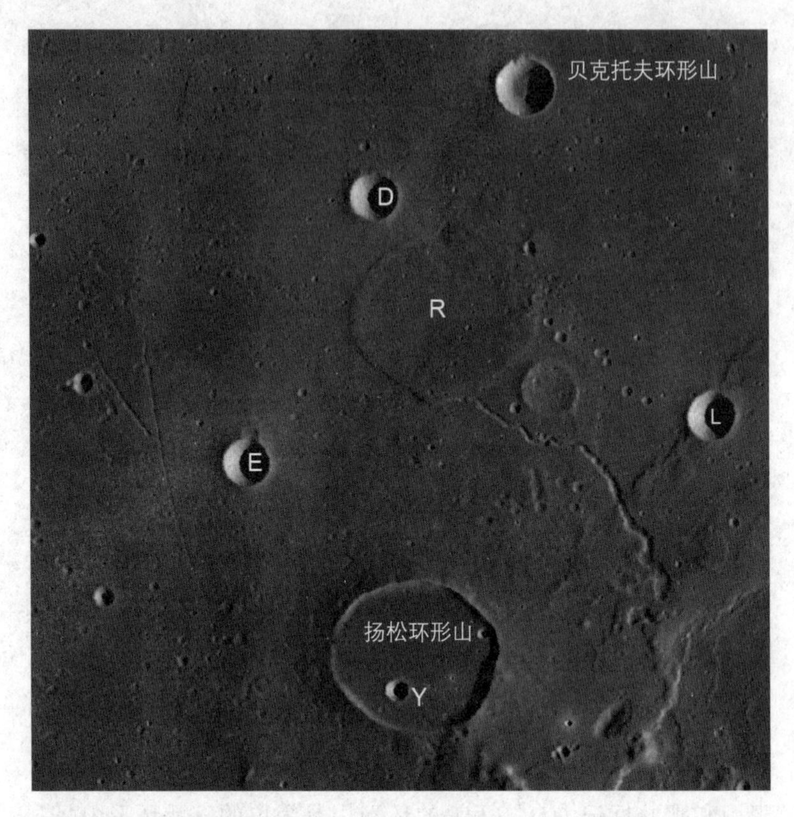

图5　静海北部110千米宽的区域，包括扬松环形山。这些字母是国际天文学联合会对一系列扬松“卫星环形山”的命名

比扬松环形山稍大一点的R环形山几乎无从辨认。你可以在正常的熔岩地面表面辨认出它的圆形轮廓。这就是所谓的“假环形山”，表示一个完全被岩浆淹没的古老环形山，几乎完全看不出来。在它的东南方向，有一个更小的假环形山（未标注字母）。

从这些简单的观察中，我们可以推断出事件的顺序如下。首先，有一次撞击形成了现在被静海所占据的大盆地。接下来，该盆地的底部出现了撞击，形成了一些环形山，比如扬松本身、扬松R以及扬松R未命名的较小邻居。然后，盆地被月海玄武岩熔岩流淹没，这可能完全掩盖了一些环形山，但不足以完全掩盖现在成为“假环形山”的可见的环形山。新的表面随后受到了形成贝克托夫、扬松D、E、L和Y以及许多较小的环形山的陨石的撞击。

月球表面形成环形山的时间尺度的依据是，月球表面存在的时间越长，其上形成的环形山就越多。这可以用我们刚才看到的相互年龄关系来检验。通过从月球带回的岩石和矿物质样本，它已经被校准为一个绝对时间刻度，这些样本已可在实验室中通过测量放射性衰变积累的产物来确定年代。它们表明，在41亿年至38亿年前的这段时间内，环形山的形成速度非常快，这段时间被称为“后期重轰炸期”。在此之前的月球表面全都没有幸存下来，所以我们不确定更早的时候发生了什么。“后期”一词表示其是在太阳系形成的后期。事实上，它是月球历史的早期，月球和地球大约有45亿年的历史。

在后期重轰炸期中形成了30个已知盆地。南极-艾特肯盆地是最古老的，可能形成于41亿年至40亿年前。雨海盆地形成于38亿年前，而东海盆地（最年轻的盆地）大约有37亿年的历史。

无论如何，这些盆地形成几亿年后，最为大量的月海玄武岩熔浆爆发了出来。大多数火山喷发是在大约30亿年前结束的，但有些地区的火山喷发一直持续到大约10亿年前。在过去的35亿年间，环形山形成的速度似乎相当稳定，比后期重轰炸期时要低得多，尽管不可能排除还有时间较短的轰炸期。

环形山并没有停止形成。正如前面提到的，第谷环形山一条“射线”中的物质（也就是第谷环形山本身）的年龄被确定为1亿年。而月球轨道飞行器已经拍摄到更小的、全新的环形山，包括一个18米宽的环形山，它的形成被2013年3月18日地面望远镜记录的短暂闪光标记。造成这个环形山的撞击物直径大约有1米，它如果撞击地球的话并不会留下环形山，因为它会在大气层中因摩擦而蒸发。

环形山与月海的命名

月球环形山的命名现在由国际天文学联合会负责。其中大多数都是以已故的科学家和极地探险家的名字命名的。一些最近命名的环形山是为了纪念那些失去生命的宇航员。月海的名字大多是描述天气状况的拉丁词语（例如，“Mare Imbrium”的意思是“雨海”）。一个例外是莫斯科海，它是在“月球3号”探测器的第一个月球背面图像中被发现并被命名的。另一个是东海，这个名字对于它位于月球西半球中央经度的特征来说似乎有些不恰当。在天平动合适时，可以在地球上看到它的一部分。这个名字是从一个天文学家使用地球天空的方向，而不是从月球表面的方向考虑坐标的时代继承下来的。

考虑到木星卫星命名的历史，你可能不会惊讶于早期的观察者对月球命名提出了相互矛盾的方案。今天的方案源于一个名叫乔瓦尼·利奇奥里（1598—1671）的意大利耶稣会士，于1651年发表。用字母来识别一个较大的命名环形山内部的环形山或周围的较小环形山的惯例（如图4中的哥白尼B和图5中的扬松D、E、L、R和Y）是由德国人约翰·马德勒（1794—1874）

发明的。注意，这些字母只表示距离。扬松D、E、L和Y一定是在扬松环形山形成很久以后才形成的，所以它们的起源肯定和扬松环形山没有关系，也很可能彼此之间也没有关系。 35

月海玄武岩和月壤

月海玄武岩似乎是在月球地幔部分温度高到足以开始“部分熔融”时形成的，这使得部分融化物被蒸发掉，在地幔中留下了固体残留物。到达地表的岩浆成分与玄武岩相似。它的黏度比大多数熔融岩石都要低，并且在固化之前能够在地面漫延很远。

火山喷发的早期阶段可能非常猛烈，熔岩在落下之前向天空喷射出数百米。喷发出熔岩的主要裂缝是看不见的，因为它们被后来更温和的渗出物所填满并覆盖，但NASA的“圣杯号”任务在2012年的重力制图中发现了一些被致密固化熔岩填满的埋藏裂缝（表2）。一些后来的熔岩流在早期熔岩流的表面雕刻出了通道，图5中有一个很好的例子：距离扬松环形山东部边缘30千米，向扬松R方向蜿蜒而去，扬松E以西，你可以看到两条又直又窄的裂缝，坐落在岩脉侵入物或者是熔岩冷却收缩时形成的裂缝上。

尽管把月球描述为一个岩石天体是正确的，但实际上月球表面只有很少的固体基岩。各种大小的撞击形成了一层碎片（估计在月陆上约为10米厚，在较年轻的月海表面约为5米厚），这些碎片构成了月球表土，更确切地说，它们被称为“月壤”。 36 大部分月壤是由小于十分之一毫米大小的碎片组成的，尽管有撞击大到足以穿透基岩，此时会抛出一些大块岩石碎片，就像图6左边的例子。

表2　月球探索中的亮点

名　　称	国家	日　　期	成　　就
月球1号	苏联	1959年1月4日	飞掠月球，没有拍下图像
月球2号	苏联	1959年9月13日	在月球表面硬着陆
月球3号	苏联	1959年10月6日	飞掠月球，首次拍摄月球背面图像
徘徊者7号	美国	1964年7月31日	在月球表面硬着陆
月球9号	苏联	1966年2月3日—2月6日	在月球表面软着陆，拍摄第一张月球表面照片
月球10号	苏联	1966年4月3日—5月30日	第一个环绕月球的飞行器
勘测者1号	美国	1966年6月2日—1967年1月7日	在月球表面软着陆
月球轨道器1号	美国	1966年6月14日—1967年10月29日	环绕月球的探测器
阿波罗8号	美国	1968年12月24日—12月27日	人类第一次绕月球航行的太空任务
阿波罗11号	美国	1969年7月20日—7月21日	人类第一次登上月球，带回21.5千克样品
阿波罗12号、14号到17号	美国	1969年11月—1972年12月	载人登月，带回总共360千克样品
月球16号	苏联	1970年9月20日—9月24日	第一个实现在月球上自动取样并送回地球的探测器，带回0.1千克样品
月球车1号	苏联	1970年11月17日—1971年9月14日	第一个无人月球车，在月球上行走了11.5千米

续　表

名　称	国家	日　期	成　就
月球20号、24号	苏联	1972年2月，1976年8月	自动取样探测器，带回175千克样品
月球车2号	苏联	1973年1月15日—5月11日	无人月球车，在月球上行走了40千米
飞天号探测器	日本	1990年3月—1993年4月	绕月轨道飞行并最终硬着陆
克莱门汀号	美国	1994年2月—6月	绕月轨道飞行器
月球勘探者号	美国	1998年1月—1999年7月	绕月轨道飞行并最终硬着陆
智慧1号	欧盟	2004年11月—2005年9月	绕月轨道飞行器
月亮女神号探测器	日本	2007年10月—2009年1月	绕月轨道飞行并最终硬着陆
嫦娥1号	中国	2007年11月—2009年5月	绕月轨道飞行并最终硬着陆
月船1号	印度	2008年11月—2009年8月	绕月轨道飞行并最终硬着陆
月球勘测轨道飞行器	美国	2009年6月至今	绕月轨道飞行器
月球陨坑观测和遥感卫星	美国	2009年10月9日	在月球表面硬着陆
圣杯号月球探测器	美国	2012年1月—12月	绕月轨道飞行并绘制月球重力场分布图
嫦娥3号	中国	2013年12月14日至今	绕月轨道飞行，软着陆并投放“玉兔号”月球车

图6 “阿波罗15号”宇航员詹姆斯·欧文和月球漫游车在哈德利沟纹(一个1.5千米宽,300米深的熔岩通道)的边缘。拍摄于1971年7月。注意前景中月壤上的脚印

月壤主要由尘埃构成,这意味着在其上留下的脚印非常清晰,事实上,阿波罗任务中的宇航员在1969年至1972年期间留下的足迹,今天还可以从月球轨道拍摄的超高分辨率图像中看到。

月球探索

在1969年7月美国“阿波罗11号”登陆月球赢下“太空竞赛”之前,月球一直是人类探索的主要目标。20世纪70年代的阿波罗计划和苏联的无人月球采样返回计划(三个着陆器一共带回了将近三分之一公斤的月壤)结束后,又过了将近20年,探索才重新开始,配备了新一代无人轨道飞行器用于特定的科学观测。更多的国家也加入进来,如表2所示。在未来,印度计划

39

在2016年至2017年发射名为“月船2号”的轨道飞行器/着陆器/月球车套件，中国计划在2017年发射名为“嫦娥5号”的样本返回任务。2014年，英国启动了一项名为“月球任务1号”的雄心勃勃的众筹项目，计划在2024年发射一枚探测器，在南极-艾特肯盆地的底面上钻孔。目前还不清楚人类何时会重返月球，但如果要我打赌的话，那将是中国在2020年左右的一次登月任务。

“别忘了带些石头回来！”这是一幅报纸漫画上的笑话，漫画中有人在发射前向“阿波罗11号”机组人员挥手告别。将月球样本带回地球，进行详细而精确的分析，是该项目最重要的科学目标。我已在前文提到了放射性年代测定法，其他的分析包括确定月球岩石的矿物成分，以及这些矿物中不同元素的丰度是否与我们在地球岩石中发现的不同。对于像氧这样有着不止一种稳定同位素的元素，这些同位素的相对丰度可以被用作“指纹”，以确定地球和月球是否由同一来源的物质构成。即使是像用显微镜观察岩石以研究其纹理这样的基本工作，也能告诉我们岩石的起源和历史，而这些是无法在月球轨道上分析出来的。

六次阿波罗着陆共收集了382千克的样本，而三次月球机器人任务又成功带回了0.32千克的样本。所有这些都是来自月球近地面的，所以一些重要的地区仍未被采样过。南极-艾特肯盆地也许是未来采样返回的最引人注目的目标。

然而，在地球上可获得的月球岩石比人为带回的要多。小行星对月球的撞击可以使一部分喷出物以足够的速度逃离月球，最终可能会以陨石的形式落在地球上。月球陨石于1982年首次被发现。它们与来自小行星的更常见的陨石类型不同，但

40

如果我们之前没有已知的月球样本与其比较，它们的起源可能仍然不明。目前作为陨石被收集起来的月球岩石大约有48千克。其中大约一半可能来自远地侧，但当然，我们无法确定任何样本的具体来源地点。

一些月球岩石是角砾岩，由在撞击中破碎的玄武岩块和/或月陆岩石块组成，同时被撞击产生的热量焊接在一起。其他的样品是单一的玄武岩或月陆岩石。月壤中含有所有这些的碎片或粉末残留，周围还有直径约为0.1毫米的类似玻璃的圆珠，似乎是火山爆炸性喷发时玄武岩喷雾冻结后的液滴。

雨海及其周围的大多数月海玄武岩的地球化学特征表明，它们来自一个钾（K）、稀土（REEs）和磷（P）相对富集的地区，被称为“KREEP”。人们认为这记录了近地侧地幔中的一块异常，其中钾的放射性衰变产生的热量（还有钍，它是通过探测其发射的 γ 射线在轨道上测量的）是一个关键的原因，造成了部分熔融，导致了月海玄武岩的喷发。

根据当时的分析技术，月球样品的一个早期特征是矿物中缺乏水分，也没有任何化学变化。有数十亿年历史的矿物晶体看起来就像去年才长出来一样新鲜，而在地球上，潮湿的环境会导致化学变化沿着裂缝和解理面渗透到矿物中。

月球是一个非常干燥的地方。除了缺水之外，它的“挥发

性元素”（如钠）的存量要低得多。另一方面，月球岩石中三种稳定的氧同位素的相对丰度与我们在地球上发现的几乎完美匹配，这表明这两个天体是由相同的物质来源形成的。然而，如果是这样的话，为什么月球没有一个与地核一样的铁核呢？一方面是较小的核心和挥发性元素的耗尽，另一方面是类似的氧同

位素丰度，这些明显的矛盾在试图解释月球最初是如何形成的时候必须加以解决。

月球的起源

1879年，乔治·达尔文（1845—1912），即相比之下更为著名的查尔斯·达尔文的次子，提出月球是由先前快速旋转的地球分裂而成的。这可以解释月球缺乏与地球相同的地核，同时氧同位素丰度相同，但不能解释其挥发性元素的消耗。而在共同吸积模型中，地球和月球在太阳系形成的过程中并肩成长，但该模型由于月球缺乏地核和不匹配的挥发性元素而失败。而如果在月球独立形成之后，其被地球捕获在动力学上非常困难，并且还需要解释氧同位素的匹配，这只能认为其是一个偶然事件。

月球的起源仍是一个有争议的问题。20世纪80年代中期，一种理论认为它是由对早期地球的“巨大撞击”形成的，并已被广泛接受，因为它可以解释这两个天体之间的相同点和不同点。类地行星成长的最后一个阶段可能是大小大致相似的天体（被称为行星胚胎）之间的一系列碰撞（巨大碰撞），而不是由许多小得多的天体之间的多次碰撞而形成一个更大的天体。巨大碰撞通常会导致两个行星胚胎的合并，由于撞击产生的热量，一个更大的行星胚胎预计将大部分熔化。熔化导致了铁（密度较大）更容易向内下沉形成核心；这个过程被称为行星分化。

42

根据月球起源的巨大撞击假说，原始地球经历的最后一次巨大撞击是一个火星大小的分化后的行星胚胎对它的一次侧边撞击。撞击并没有导致完全的合并，而是将撞击者的地幔和地球的部分地幔抛射到太空中，而撞击者的地核则向地球内推进并与地核合并。然后，月球在地球轨道上由撞击物的地幔和撞击目标的地幔混合而成。

随着月球的增长，能量转化为热量，将其融化，使少量的游离铁下沉，形成一个微小的核心。更重要的是，随着遍布月球全球的岩浆海洋开始冷却，第一批形成的晶体应该是钙长石，据探测，这种矿物构成了月球大部分的月陆。钙长石的密度相对较低，晶体更容易上升。它们可能需要聚集在一起，形成更大的物块，然后它们的浮力才能克服岩浆的黏性，随后它们会上升，直到它们到达月球表面，合并形成月陆地壳。

放射性年代测定法已经确定，最古老的月球地壳样本形成于约43.5亿年前，大约是太阳系诞生后的2亿年后，但形成月球的那次撞击可能早在45亿年前就发生了。

水和单独的挥发性元素会先在巨大撞击产生的碎片能够合并成月球之前流失到太空中，所以巨大撞击假说是最可信的。这个巨大的撞击物甚至被赋予了一个名字——忒伊亚，以神话中的塞勒涅的母亲命名。该假说最近的一种变体被用来解释近地侧和远地侧之间的一些差异，即碎片最初形成了两个卫星，在巨大撞击后的几千万年内，它们在相对缓慢的相互碰撞中合并。

43

月亮上的水

根据对阿波罗带回的数据的原始解读，月球并不像它看起来的那样完全干燥。21世纪，人们在月球样品中发现了一种名为磷灰石的矿物，其中有水的痕迹。磷灰石的晶体结构使它能吸收任何可用的水分子，以及任何大型负离子，如氟和氯形成的负离子。关于如何使用磷灰石来解释月球地幔中的水分含量，人们一直在争论不休。它可以是百万分之一到千分之一。我们也不确定地球地幔的平均含水量是多少，但几乎可以肯定它比月球的要高，可能达到大约1%。因此，月球表面的岩石和它的内部仍然被认为比地球干燥得多。

然而，月球上还有另一个不相关的能够储存水的地方。它以冰的形式出现在寒冷的零下170摄氏度的两极附近的陨击坑里，靠近两极的陨击坑底部从来没有被太阳照亮过。这方面的证据积累得很慢。1994年，克莱门汀轨道飞行器显示，无线电波从这些陨击坑反射回来的方式与冰一致。5年后，月球探勘者轨道飞行器上的中子谱仪显示了氢的浓度，最合理的解释是氢存在于水中（H_2O）。决定性的论据在2009年，当时运载月球轨道勘测器的半人马座火箭撞上了南极附近一个名为凯布斯的陨击坑的永久阴影底部。一颗名为LCROSS（月球陨击坑观测和传感卫星）的探测器在其后6分钟也坠毁了。在它坠毁之前，它能够分析由半人马座撞击产生的喷出物，结果显示，按质量计算，凯布斯的底部约有6%是水。

44

对于月球的极地冰有一个相当简单的解释，它与月球自诞生以来就存在的月球内部的水没有任何关系。当彗星撞击月球并形成陨击坑时，彗星上的冰就会蒸发，水分子加入到月球的外逸层中。如果一个分子碰到热的表面，它就会反弹，最终消失在太空中。然而，如果它碰到冷的表面，它就会粘在那里。如果它撞击到一个长期寒冷的地方，它将无限期停留在那里。这样，永久无光的陨击坑地面就像“冷阱”一样，在那里冰分子一个分子一个分子地积累，直到今天仍然如此。同样的过程也会在水星的极地陨击坑中形成冰。

45

第三章

月球对我们的影响

天空中的月球是我们最初进入太空的动力，但长期以来，它以许多其他方式渗透到人类文化中，包括为音乐和歌曲提供主题（有好有坏），并影响了语言。例如，“盈”和“亏”指的是月球被照亮的面在不同阶段变大或变小。

在我们计时的方式上，月球的影响几乎和太阳一样大。月球绕地球公转一圈需要27.3天，但由于地球在这段时间里绕太阳公转了十二分之一圈，所以太阳—地球—月球的关系需要29.5天才能完成一个周期。这个29.5天的周期是几乎所有已知的人类历法中使用的月份的起源。在英语和日耳曼语中，“Moon”和“month”（“月球”和“月份”）甚至有同一个词根。把阴历的一个月分成四个部分（例如，第二部分是上弦月和满月之间）可能就是我们一周七天的起源。

西方文化群体，以及全世界出于官方用途，都使用阳历。它通过地球围绕太阳运行的轨道上的一个固定位置来定义一年的开始。为了简洁起见，一年的365天或366天里有12个月，其中

46

除了2月以外的所有月份都比阴历月份略长。然而，伊斯兰历法是阴历，以每年为354天或355天（即12个29.5天的阴历月）来计算时间，因此这比地球绕太阳公转的周期要短。传统的中国农历和相关的东方历法都是阴阳历，使用阴历月份，但将新年定义为冬至后的第二个（有时是第三个）新月，这意味着每年的长度不同，但平均起来等于地球的轨道周期。

“难得一见的蓝月亮”这个说法最初指的是罕见的甚至是不可能发生的事情。最近，历法学家将“蓝月亮”一词正式定义出来，指的是同一月份内出现第二次满月的情况，平均每19年会出现7次这种情况。

尽管每个人应该都习惯了在天空中看到月亮，但还是有一个令人惊讶的常见错误。想象一弯新月，或者让一个毫无戒心的朋友或孩子画一个更好。他们很可能会想出一个新月形的月亮符号☾，而不是它的镜像☽。即使他们生活在热带北部，那里只有残月看起来才像一个☾，且只能在黎明的天空中看到，他们也会这样做。我不知道这是为什么。我们大多数人更经常在傍晚的天空中看到新月，而从北半球看，它的形状是☽。

从南半球看，天空中的物体看起来是“倒过来的”，在那里，傍晚天空中渐盈的月亮看起来像一个“☾”，而早晨天空中渐亏的月亮则相反。我所描述的“☾”和“☽”形状是有角度的，其中发光的弧线是向下倾斜的（朝向太阳，靠近或低于地平线），而且越靠近热带，倾斜就越大。从赤道附近的地点看，如果你在黑暗的天空中看到新月，它看起来总是像“◡”（从来不会像“◠”），因为被照亮的边缘必须面对太阳，而太阳在地平线以下，要么刚刚落下，要么即将升起。

47

海洋潮汐

海洋中的潮汐是由月球和太阳对海水的引力引起的，而海水可以做出比固体地面更自由的反应。月球的质量比太阳小得多，但它与地球的相对距离弥补了这一点，导致月球施加的潮汐力刚好是太阳潮汐力的两倍多。

月球和太阳在新月时排成一线，此时一天中有两次涨潮和两次退潮。它们将发生在接近中午和午夜时，除非海岸线使海水的运动复杂化（例如围绕不列颠群岛的潮汐）。重要的一点是，每天有两次潮汐，而不是一次，因为固体地球被拉向月球和太阳，同时被拉离留在地球夜晚面的海水。在满月时的潮汐效应是完全相同的，因为尽管月亮和太阳在向相反的方向拉扯，但是潮汐的大小是由它们对地球不同表面的拉力强度的差异决定的。在这种时候，由于太阳和月球造成的潮汐叠加在一起，潮差（潮起潮落相差的量）是最大的。这种情况被称为“大潮”，指的是潮水的快速上涨，与季节无关。相反，当太阳和月球在天空中处于直角（上弦月和下弦月）时，较弱的太阳潮汐与月球潮汐有6小时不同步，从而减小了潮差。这种情况被称为“小潮”。

48

图7以图表形式总结了这些效果。地球和月球这两个物体的固体部分也会受到同样的力的扭曲，但其扭曲程度要比海洋的扭曲程度小得多。

如果你对海边度假时的潮汐很熟悉，你可能会惊讶地发现深海的潮差只有几十厘米。海岸线能使潮差大为扩大。例如，英国东海岸大部分地区的平均潮差约为4米，但在面向大西洋的“漏斗”处（如布里斯托尔海峡和圣马洛湾），潮差可达到它的三

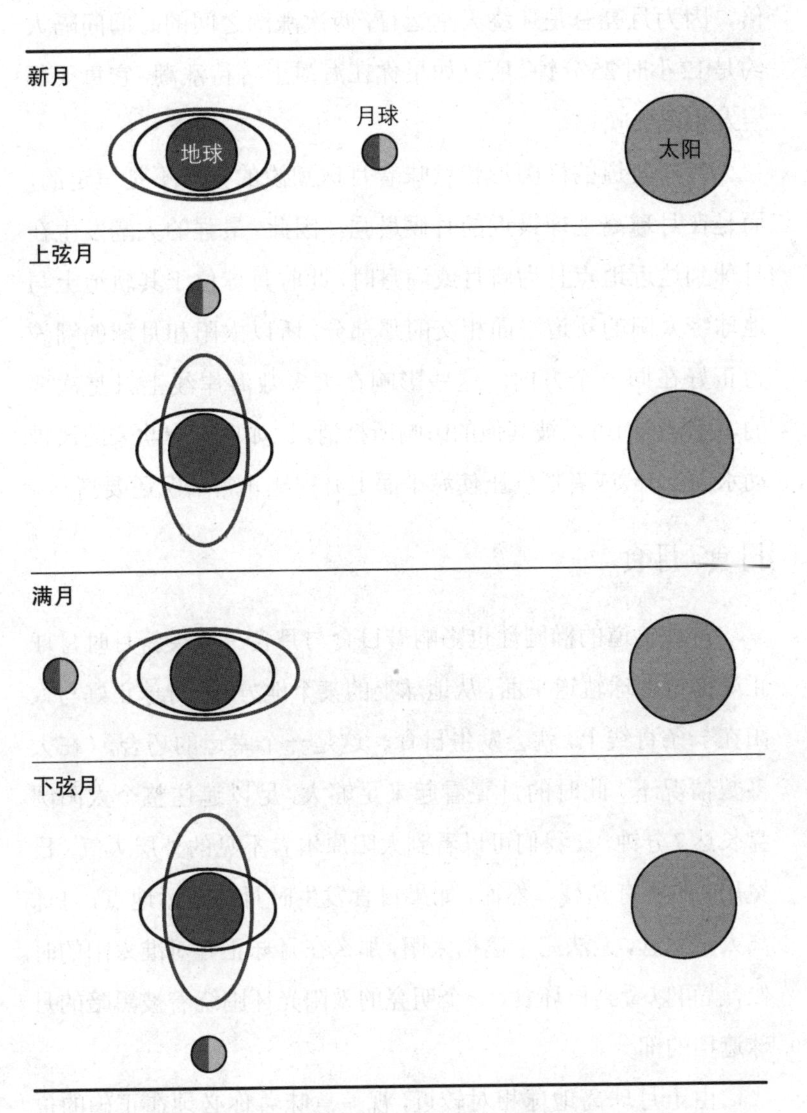

图7　太阳、地球和月球（不按比例尺）的连续四次海洋潮汐的示意图。围绕地球的椭圆显示了来自月球（较大的椭圆）和太阳（较小的椭圆）的引潮力的强度。这种影响在新月和满月时叠加，产生了最高的潮差（大潮），而在上弦月和下弦月，太阳和月球造成的潮汐相差90度，导致了较低的潮差（小潮）

倍。因为月亮总是环绕天空运行，两次涨潮之间的时间间隔大约是12小时25分钟，所以如果你在海滩上等待涨潮，它每天都会发生得更晚。

月球轨道的椭圆形状意味着月球潮汐的强度不是恒定的，而是在月球离地球最近的时候最强。因此，最强的大潮发生在月球到达近地点且为新月或满月时，此时月球位于其轨道上与地球绕太阳的轨道平面相交的那部分，所以太阳和月球的潮汐力正好在同一个方向。这些影响在大多数海岸线上只是次要的，因为它们可以被其他的影响所抵消，比如向海岸吹来的风推动水面上升，或者低气压使海平面上升得比正常情况还要高。

日食/月食

月球轨道的椭圆性也影响着日食与月食。如果新月时月球正好穿过地球轨道平面，从地球上的某个地方看，月球正好与太阳在一条直线上，就会发生日食。这是一个幸运的巧合，（在大多数情况下）此时的月亮看起来足够大，足以遮住整个太阳圆盘长达7分钟，让我们可以看到太阳原本看不见的外层大气（日冕层）的所有光辉。然而，如果日食发生时月球在远地点，月球离太阳太远，无法完全遮挡太阳，那么在月球正好对准太阳的时候，就可以看到日环食，一个明亮的太阳光环围绕着被黑暗的月球遮挡的部分。

由于月球离地球相对较近，视差意味着你必须在正确的位置才能看到日全食。即使在近地点，月球阴影在地球上经过的路径也只有250千米宽。在这条路径之外，你只能看到日偏食，此时月球只挡住了太阳的一部分。

另一方面，当月球在天空中正对着太阳时，地球投下的阴影大到足以遮住整个月球。这被称为月食，从任何月亮在地平线以上的地方都可以看到。虽然月食中的月亮是在地球的阴影下，但它仍然是可见的，因为它受到了通过地球大气层折射而变红的阳光的照射。从地球上看月球，你会看到一个暗红色的圆盘。从月球上看去，你会看到一片红色的“晚霞”环绕着地球。

轨道衰减和日长

就像地球抬升月球产生潮汐隆起一样，月球的引力也会使地球的固体部分潮汐隆起，当然还会造成海洋的潮汐。因为月球的质量要小得多，它对地球的影响要比地球对月球的影响小得多。因此，月球并没有使地球的自转速度减慢到与月球的公转周期相一致（冥王星和它最大的卫星冥卫一就发生了这种情况，但这是一个罕见的例子）。

测量结果表明，地球的自转速度每世纪都会减慢1.6毫秒，这主要是来自月球的潮汐阻力造成的。同样的力量也在减缓月球的轨道速度（与此同时，也在减缓月球的自转速度）。较慢的轨道速度要求它离地球更远，所以它正在以每年3.8厘米的速度后退。这种退行是可以计算出来的，但是也可以被证明。通过将激光束定向到“阿波罗11号”、“阿波罗14号”、“阿波罗15号”和苏联月球探测器留在月球表面的反射器上，可显示光从地球到月球的时间正逐渐变长。

如果月球现在的退行速度是恒定的，那么在10亿到20亿年前，月球离地球会非常近，不可能保持稳定，所以现在的退行速度肯定比过去更快。我们无法测量月球在遥远的过去离地球有多近，但有迹象表明当时地球的自转速度有多快。在显微镜下观察，一些珊瑚品种会显示出每日的生长线。在有3.7亿年历史的珊瑚中，这些线以每年400天为周期，表明那时每年有400天。地球的轨道周期不太可能有太大的变化，所以我们可以得出结论，为了匹配一年400天的长度，一天的长度一定更短（大约22小时）。

月球对人类行为的影响

“lunacy”（精神失常）这个词来源于拉丁语“lunaticus”，意思是一个人被认为受到月球影响而变得疯狂。然而，只要做了足够仔细的测试，就会发现月相与精神疾病或任何人类行为（包括犯罪、自杀或出生率）之间都没有关联。没有任何生理上的原因来造成任何影响，例如，人体受到的潮汐力是非常小的。

在大多数文化中，人们普遍认为月球对人类的生育能力有影响，这可能是因为一个月的长度和人类的月经周期相似。然而，这似乎是一个巧合，两者之间没有任何联系。

不过，有一个月球影响人类行为的现代例子，这就是所谓的“超级月亮”。在天空中，月球在近地点的角尺寸大约比在远地点大14%，所以如果满月时月球在近地点，那么满月会比平均水平要大一些。除非你真的去进行测量，否则它们的差别很小，根本看不出来。但是尽管如此，在过去的几年里，每当满月时月球接近近地点，社交媒体网站和传统新闻媒体上就会出现一片混乱，它们纷纷表示这一现象将会十分壮观，甚至可能引发某种自然灾害。

“超级月亮”一词是在20世纪80年代由一位占星者创造的，

用来表示当月球位于其轨道上最接近近地点的那十分之一部分时出现的满月。这个引人注目的词已经吸引了公众的想象力，并取代了更麻烦的短语“近地点满月”。然而，这是一种误导。“超级月亮”只比平均大小大一点点，甚至不是一种特别罕见的现象（它比“蓝月亮”还要常见）。大约每10个满月中就有一个符合“超级月亮”的标准，事实上，它在2014年的7月、8月和9月就连续出现了3次。我们将“超级月亮”在视尺寸上的平凡无奇和微小差异与虚构的“超人”所拥有的惊人能力或“超级火山喷发”进行对比：“超级火山爆发”指的是一场灾难性的喷发，其产生的火山灰量至少是过去1 000年里任何一座火山所能喷发出的火山灰量的10倍，在地球上的某个地方平均每5万年才会发生一次。目前还没有发现“超级月亮”和火山爆发、地震等事件之间的关联，也不太可能找到。

尽管月球近地点的角尺寸在天空中更大，但此时的月球表面实际上并没有那么亮。月球表面的亮度取决于月球到太阳的距离，而不是它到地球的距离。最主要的因素是地球轨道的偏心率，这导致地球—太阳（和月球—太阳）距离最近（1月初）和最远（7月初）之间有大约500万千米的差距。这盖过了地球—月球间距离的4万千米的变化，并导致月球表面亮度变化了7%的范围。然而，由于月亮在天空中的角尺寸在它最近的时候更大，当地球—太阳距离固定时，满月出现在近地点比满月出现在远地点时要多出30%的月光。

人眼能适应不同的亮度。我们的瞳孔扩张或收缩，以允许适量的光线进入视网膜，但我们的记忆并没有储存这些信息。因此，如果不使用仪器，我们就无法对一个满月和几个星期或几个月后的另一个满月进行定量比较，如果有人只是看着“超级月亮”，就说他们能看出它比平常更大或月光更亮，那是在自欺欺人。

与炒作起来的、几乎完全是想象出来的“超级月亮”现象相反，还有一种不相关的光学错觉，可以真正让月亮看起来很大。每当我们看到月亮低挂在天空（越靠近地平线越好），月亮就显得很大。实际上它并没有变大，如果你测量它，它的大小并不会改变。这是一种被称为“月亮错觉”的效应，似乎是大脑处理信息方式的结果。当月亮高高地挂在天空时，它似乎是渺小的、孤立的，漂浮在一片星海中。然而，当它接近地平线时，就会出现一些遥远但熟悉的物体与它联系起来，比如树木和屋顶，月亮在这些物体的后面显得非常突出。

月球与地球生命

月球在很多方面影响着野生动物的行为。夜行性陆生动物的活动会随着月相的变化而变化，这取决于满月前后额外的光线对它们作为捕食者的成功或作为猎物的安全是有利还是有害。许多海洋物种把月球作为一个时钟来触发大规模产卵，这并不是因为光照水平很重要，而是因为这种繁殖策略的成功依赖于同步。一些海龟会等待大潮，这样它们很容易就可以上岸，并在干燥的地方产卵，直到它们的幼崽孵化。

54

月球可能在更根本的方面影响了地球上的生命，而不仅仅是影响食物和繁殖习惯。有人认为，如果没有月球引起的潮汐，海洋生物要迁移到陆地上就会困难得多。这可能是错误的，因为太阳自身会引起每天两次的潮汐，其影响范围大约是月球和太阳共同引起的潮汐平均总和的一半。这将为海洋生物提供大量的机会，让它们发现自己被退潮暂时搁浅了。

月球的另一个影响是，它的存在可能会稳定地轴的倾斜。目前的倾斜角度是23.4度。地轴指向空间中的一个方向，且其倾斜角度变化得非常缓慢，而这种倾斜是地球绕轨道运行时季节变化的原因。计算表明，在过去的500万年里，倾斜的角度在约为4.1万年的周期内在2.5度的范围内变动。这影响了气候，并可以从化石记录中看到。20世纪90年代的研究表明，如果地球没有月球，我们的地轴倾斜将经历更剧烈的波动，从0度到近90度。这将导致严重的极端气候，甚至比火星所经历的更糟糕。火星只有很小的卫星，目前其地轴倾斜角度在5度到45度之间变动。

这种极端的气候使得高级生命不可能在陆地上生存，所以如果月球不存在，那么人类也就不存在了。然而，最近的研究对月球的重要性提出了质疑，并认为即使没有月球的影响，地球的地轴倾角也可能保持在狭窄的范围内。

月球基地和月球资源

当人类最终重返月球时，停留时间很可能会比每次阿波罗探测器登月时在月球停留的几天时间更长。如果目标是探索月球，那么有很多很好的科学理由来派遣训练有素的宇航员（最好是地质学家）。人类可以比遥控月球车更有效地利用时间。1973年，苏联“月球车2号”花了近4个月的时间，探索距离才超过“阿波罗17号”的宇航员总计22小时的三次步行的距离。

除了可以更多了解月球，前往月球也会促进其他科学的发展。在月球的背面，月球会保护你免受地球的无线电干扰，这使得建造月背射电望远镜成了一个对天文学家来说十分诱人的前景。奇怪的是，月球可能也是发现早期地球环境证据的最佳地点。就像对月球的撞击会释放出抛射物，其可以作为地球上的陨石被人收集，而对地球的撞击就会向月球抛出可以在月球上收集到的抛射物。如果30亿年前，一块地球表面被抛射到月球上，那么这将是一个奇妙的发现，因为与仍在地球上的任何已风化的遗迹相比，它将仍处于原始状态。我们希望能找到微小但可分析的早期地球大气样本，这些样本被困在受震激玻璃里的气泡中，或许还能找到记录地球生命最初阶段的微化石。

将资源从月球带回地球是否有经济意义是有争议的。到月球去取东西的成本非常高昂。由于月球如此干燥，如果能像地球一样，在月球上发现通过地壳中的水流循环形成的富集的矿物，那就太令人惊讶了。不过，在金属小行星撞击形成的环形山中，可能隐藏着诱人的铂族元素矿藏。

能从月球向地球出口的一种可能有意义的商品，是一种叫作“氦-3”的氦同位素。它在地球上是很罕见的，但从月球上采集的样本显示，月壤中氦-3的含量约为一亿分之一，据信是由太阳风带来的。氦-3和氢的一种重同位素（“氘”，可以从海水中提取）是核聚变反应堆所需的燃料，这被认为是一种“清洁”的电力来源。核聚变发电的技术可行性还有待证实，所以这一切都只是推测。然而，如果地球上真的出现氦-3市场，那么开采月球表面的月壤，将其加热并收集排出的氦-3气体将在商业上变得可行。

在月球上利用月球资源则是另一回事。从一个永久无光的

56

环形山中提取的水可以使附近的月球基地的运营成本更低。在一个极地环形山边缘的山顶上安装的太阳能电池板可以不间断地收集太阳能，这些太阳能可以用来加热月壤并使其形成建筑材料，或者驱动更复杂的过程来提取氧气和金属，又或制造玻璃纤维。即使是从地球带来的生存环境单元，在它们上面简单地堆积月壤以抵御太阳风也是有意义的。

谁拥有月球？

目前还没有公认的法律框架确定月球或其任何部分的所有权。联合国于1979年发起的《关于各国在月球和其他天体上活动的协议》（简称《月球协定》）宣布，月球应该为所有国家和国际社会所有人民的利益服务。它还表达了防止月球成为国际冲突来源的愿望。这些都是很好的观点，但协定是无效的。在能够独立飞向月球的国家中，只有印度签署了（尽管没有批准）该协定。

该协定还禁止任何组织或个人拥有地外财产，除非该组织是国际性政府组织。如果这真的有效，它将成为商业驱动下探索月球和太阳系其他部分的主要障碍。一个更务实的方法可能是接受这样一种观点：如果有人投资勘探，并且能够为一种资源找到市场，那么他们就应该被允许把它卖给任何想买的人——无论是地球上的还是地球外的。

2013年提交给美国国会委员会的《阿波罗月球遗产法案》规定，出于非商业动机，可以对月球的一小块区域提出所有权要求。该计划旨在保护六个阿波罗探测器着陆点的轮轨、脚印和硬件。这在我看来是合理的。这些都是人类共同遗产的一部

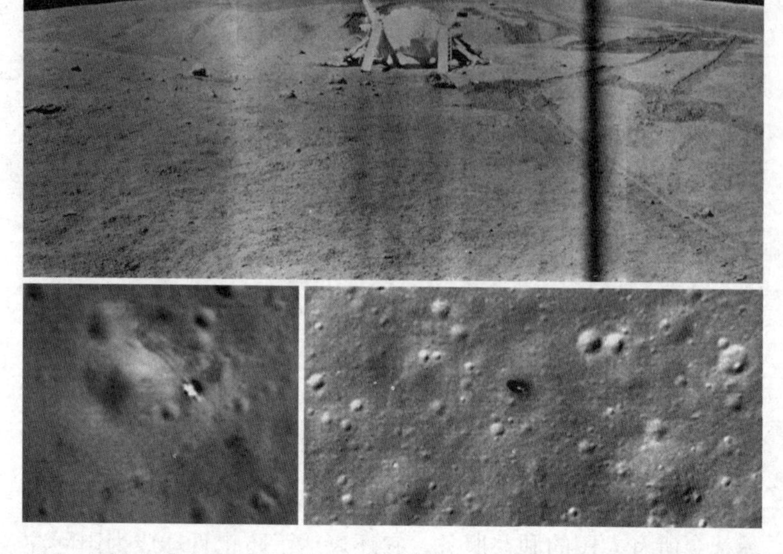

图8　月球上的私人财产？最上方是“月球21号”着陆器，1973年1月由“月球车2号”拍摄。左下角是“月球21号”，右下角是“月球车2号”，2010年和2011年由月球勘测轨道飞行器拍摄

分，如果一些“太空海盗”可以不受惩罚地破坏这些地点，或者拿走一些硬件在地球上私下出售，那将是可怕的。

58

这是一个灰色地带，但大多数立法者都会同意，已经在月球上着陆某个物体的国家或组织仍然拥有该物体，尽管不一定拥有该物体所在的地面。1993年，一个俄罗斯航空航天公司，拉沃契金协会，在苏富比的一场拍卖会上出售了该公司制造的“月球21号”着陆器（图8）和“月球车2号”。它们拍出了6.85万美元的价格，买家是英国/美国人理查德·加里奥特（1961—　），他是一名电子游戏开发者和企业家，后来自费访问了国际空间站。

59

第四章

巨行星的卫星

木星、土星、天王星和海王星等巨行星都有大量的卫星伴随。自然，第一颗被发现的卫星是最大的，通常被称为“规则卫星”，但它们只是故事的一部分。并不是所有巨行星的卫星都能被整齐地归类，但总体情况如下：

最接近行星的是内卫星，它们大多半径不到几十千米，形状不规则。它们与行星的环系统密切相关。它们的轨道是正圆形的，位于行星的赤道平面上，半径小于行星本身的3倍。

其次是半径超过200千米的大型规则卫星，大到足以用它们自身的重力把它们拉成接近球形的形状，这种情况被称为“流体静力平衡”。它们的轨道比内卫星的轨道略微更不接近正圆，半径是行星的20倍到30倍，同时也非常接近行星的赤道平面。

最后是不规则卫星，大多半径不到几十千米。这个术语既指它们形状的不规则性，也指它们轨道的不规则性。它们的轨道可能极其椭圆，通常相对于行星的赤道相当倾斜。它们的轨道半径是木星和土星半径的400倍，是天王星半径的800多倍， 60

海王星半径的近2 000倍。

除了海王星的海卫一外，内卫星和所有的规则卫星都以它们的行星自转的相同方向绕轨道运行，这被称为“顺行”运动。大多数不规则卫星都有着倾斜的轨道，都以与行星自转方向相反的方向绕轨道运行。这被描述为“逆行”运动，并暗示了这些卫星的起源。

冰

一般来说，大型卫星都不是岩石天体。在20世纪50年代，装有分光计的望远镜被用来测量物体反射阳光的特性，并将其对准了巨行星的较大卫星，然后就发现了大多数卫星表面都存在结冰的水。这并不奇怪，因为这些天体离太阳都很远，这意味着木星卫星的平均表面温度约为零下160摄氏度，土星卫星的平均表面温度为零下180摄氏度，天王星卫星的平均表面温度为零下200摄氏度，海王星卫星的平均表面温度为零下235摄氏度。在这样极低的温度下，冰的物理性质变得非常坚固，就像岩石一样；它可以支撑环形山、悬崖和山脉，而不会像地球上的冰川那样流动。

同样重要的是，当这些卫星形成时，其温度也很低。木星离太阳的距离是地球的5倍，位于“冻结线”之外。这里的温度很低，可以让围绕着年轻太阳的气体云中的水直接凝结成冰。在冻结线以外形成的天体通常含有比岩石更多的冰，因为在形成太阳系的气体和尘埃云中，形成水的所需元素（氢和氧）比形成岩石的关键成分（硅和其他金属元素加上氧）更多。在冻结线外的氢倾向于形成固体化合物，所以岩石只在冻结线内的卫星上占主导地位。

碳和氮也是常见的元素，它们组成了其他种类的冰，在离太阳更远的地方凝结。可能在土星上，当然也会在天王星和更远的星球上，冰不仅仅是由水结成的，还混合了冻结的氨（NH_3）、甲烷（CH_4）、一氧化碳（CO），甚至（在海王星）有冻结的氮气（N_2）。在巨行星中，大部分的这些物质——尤其是水，以冰的形式存在于它们的内部，位于一层主要由氢和氦组成的厚厚的气体包层之下。但卫星的重力太小，无法收集大量的气体，所以冰占主导地位。

大量的冰解释了大多数卫星的低密度。附录中的表格中，大多数巨行星的规则卫星的密度在每立方米1 000千克到2 000千克之间。岩石的密度应该大于每立方米3 000千克，而水冰的密度是每立方米1 000千克（其他冰的密度更小）。因此，卫星的体积密度越低，它包含的冰就越多，岩石越少。

卫星探测任务

如果不是太空探测器造访了这些巨行星和它们的卫星，我们接下来就没什么可说的了。探索始于飞掠（飞越行星的任务），但在木星和土星的案例中，探索已经进行到轨道旅行的阶段。这两个行星都有探测器围绕并进行了为期几年的任务，而且期间还能够至少对规则卫星进行多次近距离飞掠。近距离飞掠卫星可以实现详细的成像。探测器通常会接近到足够近的地方，以便观察卫星如何影响环绕行星的强磁场，并探测卫星是否也有自己的磁场。探测器在接近卫星时，其轨道轻微偏转的大小可以确定卫星的质量，而知道了卫星的大小，就很容易算出它

62 的密度。

故事始于1973年12月飞掠过木星的NASA的“先驱者10号”，以及在1974年12月飞掠过木星，然后在1979年9月飞掠过土星的“先驱者11号”。它们的主要目的是测量行星的大气和磁场，所以几乎没有收集到关于它们的卫星的数据。

NASA的两个“旅行者号”探测器真正打开了我们对这些卫星的视野。“旅行者1号”于1979年3月穿越木星系统，并于1980年11月穿越土星系统。2012年8月，它成为首个穿越太阳风完全消失的日球层顶并进入星际空间的太空探测器。“旅行者2号”飞掠了所有四颗巨行星：1979年7月飞掠木星，1981年8月飞掠土星，1986年1月飞掠天王星，在1989年8月飞掠海王星。它仍然是唯一一个访问过天王星和海王星的探测器。

NASA的“伽利略号”探测器于1995年12月进入木星轨道。在向木星发射了一个大气进入器后，它绕着卫星转了8年，直到它的机动推进剂耗尽，之后被允许撞向木星。因为它的主要通信天线——一个抛物面天线，未能展开，所以早期出现了严重的挫折。这意味着数据必须使用备用的“低增益”天线传输，从而减少了可以收集到的图像总数，但飞行过程中的编程和数据压缩技术挽救了很多科学研究。

NASA和欧洲航天局（ESA）的联合任务“卡西尼-惠更斯号”于2004年6月抵达土星。它在2005年1月释放了降落在土卫六表面的“惠更斯号”着陆器，而“卡西尼号”轨道飞行器则开始了漫长又复杂的轨道旅行，最终计划在2017年进入土星大63 气层。

2000年12月，“卡西尼号”在飞往土星的途中飞掠木星，在接下来的几天里，“卡西尼号”补充了“伽利略号”轨道飞行器对木卫一火山喷发的观测。NASA的“新视野号”任务提供了更多关于这些壮观事件的图片。2007年2月，“新视野号”在飞往冥王星的途中近距离掠过木星，并于2015年7月飞掠冥王星。

木星的规则卫星

木星的四颗伽利略卫星就是典型的规则卫星。在这里，我把它们看作同一个类别，将对单个卫星的讨论保留到第五章。它们在图9中被一起展示，被剖开以揭示它们的内部结构。这些图片中的结论主要是根据“伽利略号”飞越木星时获得的内部密度分布的线索，以及“旅行者号”和“伽利略号”对每颗卫星和木星磁场之间相互作用的测量得出的。后者表明，木星的磁场在木卫二和木卫四内部激发了一个磁场，很可能是通过内部含盐海洋的导电实现的。木卫三自身有一个相当强的磁场，这可能是由在其外核的液态硫化铁的对流产生的，就像在水星和地球内部的发电机理论一样。木卫一的磁场特征还没有得到很好的描述，我们不能确定它是由流体核的运动引起的磁场，还是一个源相对较浅的感应磁场。其中的三颗卫星是完成分化后的天体，密度较大的物质能够向内分离形成核心，但木卫四缺乏强烈的内部密度梯度，这表明它只有微弱的分化。

木卫一是太阳系中密度最大的卫星，也是唯一一颗表面没有冰的规则卫星。它可以被认为是我们的卫星即月球的一个更大（也更活跃）的版本。它有一个岩石构成的表面，被持续的火山爆发所散发的硫化物染成黄色和红色。

木卫二是一个更小版本的木卫一（也不那么活跃），表面被水覆盖，在接近表面（冰层）的地方是固态的，在深处是液态的，64形成了一个全球性的海洋。木卫二的密度几乎和月球一样大，这表明，由水（冰与液体）组成的外壳只有大约100千米厚。木卫二表面的冰能够破裂和迁移到相对内部的地方，但无论是木卫二还是其他任何卫星，都没有表现出与地球上构造板块的形成和迁移相似的行为。

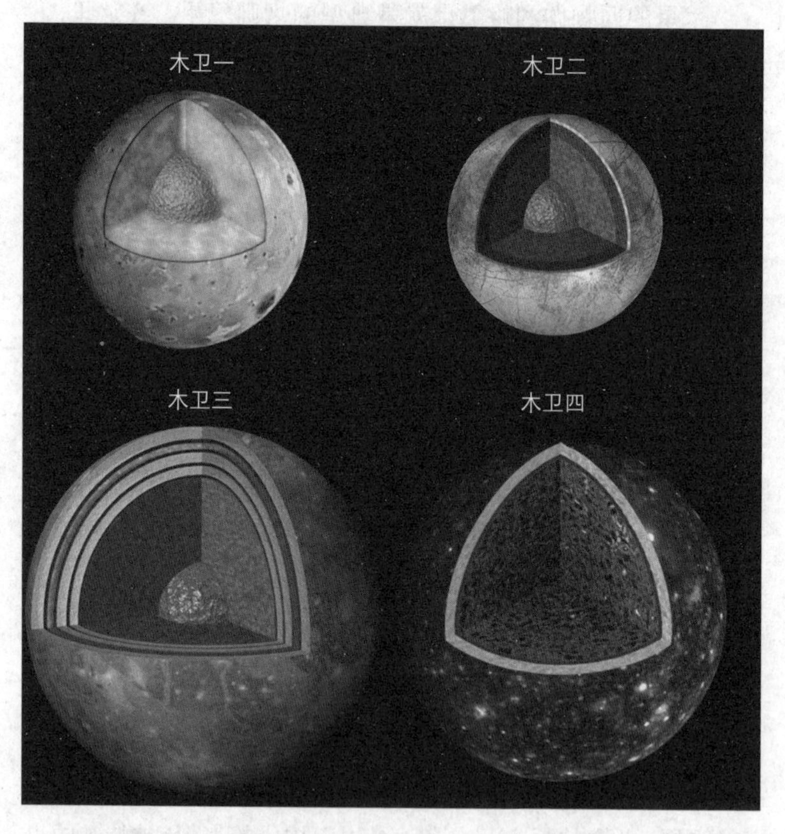

图9　剖面图显示推断出的木星规则卫星的内部结构，按比例显示。木卫一、木卫二和木卫三的核心富含铁，被岩石包围。木卫二、木卫三和木卫四的外层是冰，内部的深色层是液态水。木卫四的内部是未分化的岩石和冰的混合物

65

木卫三是太阳系中质量最大的卫星，也是唯一一颗不同的伽利略卫星。它的内部压力足以将水-冰压缩成更致密的晶体结构。它表面的冰是普通类型的冰，被称为冰Ⅰ，但在更深的地方，计算出的压力足以将其压缩成冰Ⅲ，然后是冰Ⅴ，最后是冰Ⅵ。图9显示了2014年提出的木卫三的复杂内部结构。根据这个模型，每个冰层下面都有一个液体层，形成一个多层三明治般的外观。如果这些液态层存在，它们很可能是强度非常高的盐水，含有从岩石内部溶解出来的盐，使它们在由纯水构成的冰无法融化的低温下保持液态。

伽利略卫星的密度在离木星越远的地方越小。这是一个有关它们起源的有力线索，表明它们是由环绕年轻木星周围的一团碎片云形成的（就像太阳周围的行星形成的方式一样），木星辐射出的热量足以使木卫一和木卫二缺水，而更远的卫星则能获得水。木星周围的碎片会和木星一样旋转，这自然会导致这些卫星的轨道靠近木星的赤道面，正如我们看到的那样。

轨道共振和潮汐加热

当“旅行者号”计划执行任务时，规则卫星被认为是相当沉闷的地方。有人认为，由于它们大部分都是冰，而且相对较小，任何内部岩石含有的能产生热量的放射性元素都太少，因此在过去的三四十亿年间，岩石内部不可能驱动任何活动来影响它们的表面。因此，它们将是“死亡”的世界，像月陆一样被上面的陨击坑严重破坏。岩石还是冰，这并不重要：撞击造成的环形山看起来几乎一样。

然而，事实证明木卫一的表面非常年轻，根本没有发现陨

66 击坑，而木卫二的陨击坑也很少。木卫三和木卫四上有很多陨击坑，其中有些你应该可以在图9中看到，但是木卫三的表面有大片较浅的、年轻的地形。所以，除了轨道越靠外、密度减小越多的趋势外，伽利略卫星也有轨道越靠外、表面年龄增加越多的趋势。

轨道靠近木星的卫星实际上并不年轻，但最近的地质活动使它们的表面焕然一新。原因就在它们的轨道上。长期以来，潮汐摩擦力减缓了它们的自转速度，因此它们的同步绕转与轨道周期相匹配。这也使得它们的轨道比月球更圆，因为它们围绕着一个质量更大、引力更强的行星运行。

在椭圆轨道上，天平动会使卫星的潮汐隆起在其平均位置附近来回移动。要造成卫星内部的扭曲使这种情况发生，就必然会通过内部摩擦的方式增加热量，并在它的自转变得同步之前促使其内部变得分化，如果此前更加强大的潮汐加热作用还没有使其分化的话。如果你想体验内摩擦加热的功效，可以试着将金属衣架前后弯曲，然后（小心地！）将弯曲的部分放到你的嘴唇上。

如果木星只有一颗卫星，那么木星的引力将用不到1亿年的时间迫使卫星的轨道变成一个完全的正圆形，这样就不会有天平动和潮汐加热了。然而，正如你们所知，木星有四颗规则卫星。当“旅行者1号”向木星加速时，美国人斯坦·皮尔（1937—2015）和他的同事正在计算这些卫星群的成员对潮汐加热的影响。

木卫二和木卫三的轨道周期正好是木卫一的2倍和4倍，所以木卫一每绕木星转4圈，木卫二正好绕转2圈，木卫三正好绕转1圈。这种情况被称为“轨道共振”，并且是由卫星之间的相互引力作用造成的。这意味着木卫一每次经过木卫二时的轨道位置都是相同的，木卫二经过木卫三时的轨道位置也是相同的。卫星之间反复的轻微引力拉扯使得它们的轨道无法完全变成正圆形。如果你将它们画在纸上，它们看起来就像正圆形，但它们与正圆形的区别足够大，轨道速度会发生变化，所以潮汐隆起会稍微来回移动，高度也会膨胀和收缩，从而加热它们的内部。

皮尔和他的同事计算了这一过程在木卫一内部产生的热量。在对木卫一的内部结构和强度做了最好的假设后，他们在1979年3月2日的《科学》杂志上发表了他们的发现，写道：“木卫一的潮汐能量的耗散很可能融化了它的的大部分质量。这造成在‘旅行者1号’即将传回的木卫一表面的照片中，将可以明显看出其内部大部分是熔融的。”这是我所知的行星科学中最著名的掌握时机的例子：六天后“旅行者1号”记录的图像显示木卫一上有喷发的火山，而且完全没有现存的陨击坑。

轨道共振是一种复杂的情况。当处于共振状态时，受迫的偏心量（以及因此造成的潮汐加热的速率）可以增加或减少，而且卫星可以在数百万年的时间里漂移到共振状态之外。现在的潮汐加热造成了木卫二年轻的表面，而过去的潮汐加热可以在木卫三和其他几颗巨行星的规则卫星上发现。

其他规则卫星

其他巨行星的规则卫星群缺乏在木星所发现的“轨道越向外，密度减小越多”的趋势，而除了土星的土卫六外，其他的卫星

68 就更小了。然而，土星和天王星的卫星轨道特征是相似的；在其行星的赤道平面上的卫星的轨道几乎是正圆形的，甚至在天王星也是如此：很久以前，天王星发生了一场大灾难，将整个行星与卫星系统倾斜到了轨道的一边（天王星的自转轴相对于它绕太阳的轨道倾斜了97.9度）。目前，土星的一些卫星之间有轨道共振，但天王星的卫星之间没有，尽管一些卫星的表面有潮汐加热的迹象，表明共振在过去发生过。

土星有七颗规则卫星（见附录）。从里向外分别是土卫一、土卫二、土卫三、土卫四、土卫五、土卫六和土卫八。土卫一与土卫三的共振率是2∶1，尽管这似乎没有产生有效的潮汐加热。土卫二与土卫四的共振率是2∶1，这必然为“卡西尼号”在土卫二南极附近发现的爆发羽状物提供了动力，这将在第五章中进行讨论。

天王星有五个规则卫星：天卫五、天卫一、天卫二、天卫三和天卫四。在这些卫星中，天卫一和天卫三的表面被巨大的裂缝贯穿，这最可能归因于古代的潮汐加热。天卫五有一个复杂的表面历史，可能是之前与天卫二3∶1轨道共振的结果。它是略微非球体的：它的三个轴的半径分别为240千米、234.2千米和232.9千米，有些人可能认为它是天王星最外层的内卫星。

这些卫星都可能以类似于木星的伽利略卫星的方式围绕它们的行星形成，但海王星最大的卫星海卫一（Triton）就不是这样了［不要与土星上名字相似的土卫六（Titan）相混淆］。它有一个倾斜和逆行的轨道。目前还不清楚海卫一是如何在海王星附近形成并最终进入这样的轨道的，所以它很可能是从其他地69 方被海王星捕获的。

最有可能的情况是，海卫一开始属于“柯伊伯带”，后者是一个冰天体群的统称，它们的大小在1 500千米以下，在海王星轨道附近或更远的地方绕太阳运行。如果这是正确的，那么海卫一最初独立环绕太阳的轨道一定有一次使它离海王星足够近而被捕获。捕获是很难实现的，因为要抵消太多的动量。通常情况下，接近的天体会掠过行星，或者与它相撞（十分少见）。然而，如果接近的物体实际上是两个（一个主要天体和一颗卫星），其中一个可以被捕获，而另一个将以比它到达时更快的速度被甩出去，而系统的总动量是守恒的。

事实上，今天在柯伊伯带就可以看到这样的双重天体。冥王星及其巨大的卫星冥卫一就是一个典型的例子。海卫一比冥王星稍大一些，但我们不知道它是到达海王星的那个双子星系统中更大的部分还是更小的部分。不管是双星还是单星，海卫一的到来和捕获将会打乱任何已有的规则卫星，这些卫星一定是在柯伊伯带丢失或在相互碰撞中被摧毁了。

特洛伊卫星

当一个较小的天体围绕一个较大的天体运行时，存在两个稳定点，一个更小的物体可以停留在这两个稳定点上，或者它可以围绕这两个稳定点振荡。两种情况分别发生在绕轨道运行的较小天体前方60度和后方60度，围绕较大天体共享同一轨道。从技术上讲，这些点被称为拉格朗日点，分别表示为L4（前）点和L5（后）点。然而，它们通常被称为“特洛伊点”，因为有一组在木星围绕太阳的轨道上的小行星被以特洛伊战争中的人物命名，这些特洛伊小行星聚集在木星的L4点和L5点周围。

土星有四颗小卫星与两颗规则卫星（土卫三和土卫四）处于前方和后方的特洛伊关系。半径为10千米到20千米的土卫十四、土卫十三和土卫十二是在20世纪80年代至90年代用望远镜发现的。最小的是土卫三十四，直径不到3千米，是在2004年“卡西尼号”拍摄的图像上发现的。

70

来自“卡西尼号”的图像（图10）揭示了这些特洛伊卫星表面一些令人惊讶的特征。它们表面的环形山相当少，所以可能相对年轻（这可能仍然意味着它们有超过10亿年的历史）。图

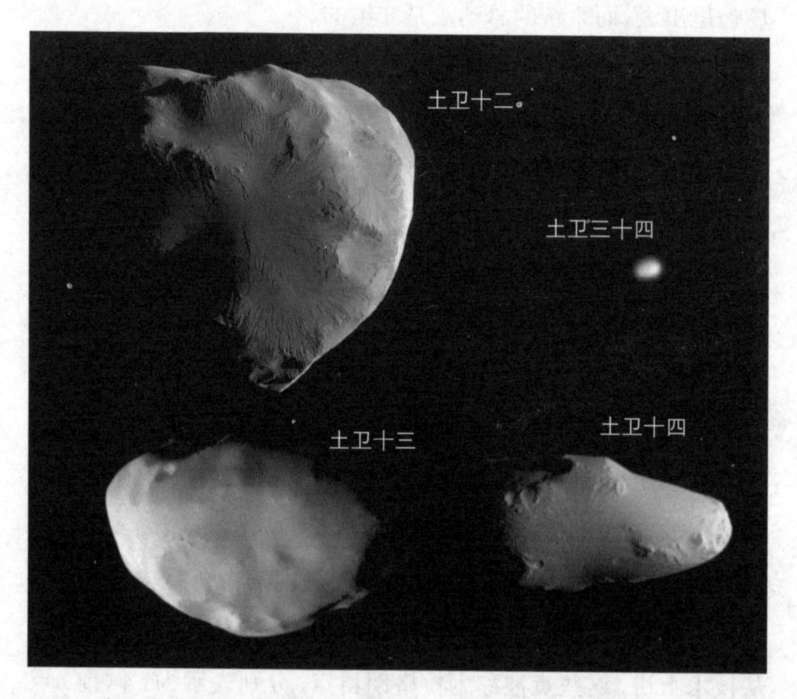

图10　太阳系中已知的四颗特洛伊卫星，由“卡西尼号”拍摄，并以近似正确的相对比例显示。土卫十二的平均半径为17千米。因为从未在足够近的距离观察过土卫三十四，所以人们不清楚它的表面细节。土卫十二和土卫三十四是土卫四的共用轨道的特洛伊卫星，而土卫十三和土卫十四是土卫三的共用轨道的特洛伊卫星

中显示的是土卫十二背向土星和土星环的一面；面向土星的那一面有更多的环形山。土卫十四的表面是整个太阳系中反射性最强的，但这在图10中并不明显，因为图像经过处理后，每个表面都同样清晰。这可能是土卫十四扫过了土卫二喷发的冰晶的结果。土卫十四和土卫十二的表面上都显示出奇特的沟壑。在它们的表面条件下，不可能有稳定的液体。某种干燥的雪崩过程可能是原因之一，但对于一个表面引力仅为地球引力的0.02%的天体来说，这要如何起作用仍是一个谜。

71

土星是已知的唯一拥有特洛伊卫星的行星，但这可能是观测偏差的结果。伽利略轨道飞行器并没有在木星上寻找它们，那里可能有像土卫三十四这样的小行星。“旅行者2号”没有机会在天王星和海王星上寻找任何卫星，在那里，即使是土卫十四大小的特洛伊卫星也很难从地球上用望远镜观测到。

不规则卫星

已知的不规则卫星比其他任何类型的卫星都要多。木星就有59颗。其中最靠内的7颗在顺行轨道上，最大轨道半径为木星半径的238倍（木星的赤道半径为71 492千米），其他的都在外部的的逆行轨道上，最大半径为木星半径的400倍。这是由于不同轨道的长期稳定性相对于行星的希尔球范围（行星的引力超过太阳引力的范围）不同的结果。在几十亿年的时间里，顺行轨道只在希尔球半径的一半范围内是稳定的，而逆行轨道可以稳定在希尔球半径的三分之二范围内。木星的希尔球半径大约是木星半径的740倍。

木星最大的也是离木星第三近的不规则卫星——木卫六，

半径为85千米，早在1904年就被发现了，但大部分卫星都是自2000年以来通过专门的望远镜观测发现的。木星最小的卫星半径只有1千米。即使是木卫六在“伽利略号”最好的图像中也只有7个像素，所以我们对它们的了解很少。

另外三颗不规则顺行卫星的轨道与木卫六相似（半径约为

木星半径的160倍）。这四颗卫星都能以与碳质小行星的原理相同的方式反射阳光，所以这四颗卫星都被认为是一颗碳质小行星的碎片，它们被木星捕获后解体。

在木星逆行的不规则卫星中，可以识别出三组由共同的反射率特征和相似的轨道半径、偏心率和倾角定义的卫星。它们都以它们中最大的那个成员来命名：7颗卫星组成的阿南克（即木卫十二）群的轨道接近木星半径的297倍；13颗卫星组成的加尔尼（木卫十一）群的轨道接近木星半径的327倍；7颗卫星组成的帕西法厄（即木卫八）群的轨道接近木星半径的330倍。这些被认为是其他类型的小行星的碎片。许多木星的不规则卫星，包括帕西法厄群外围的20多颗，都没有已知的群，它们中的每一个都可能是被捕获的小行星或彗星核。

尚不清楚这些卫星是何时被捕获到的。在木星形成后的短时间内会更容易实现捕获，因为此时木星有一个扩展的弥漫大气层，提供了必要的阻力，使进入其中的物体足够慢，从而被捕获。同样，尚不清楚每个相关卫星群的源头是何时破碎的。这可能发生在捕获过程中，也可能发生在碰撞之后。

木星的不规则卫星没有一个表现出潮汐锁定。它们太小，离木星太远，潮汐力无法发挥作用，这一点在少数几个例子中得到了证实，它们的自转周期被测出只有几个小时，而公转周期则有数百天。另一方面，它们受到太阳的强烈引力，以至于它们轨道的形状和倾角在短短几年内就会发生显著变化。

其他巨行星的不规则卫星也遵循与木星相似的模式，它们的起源也可能相似。最遥远的卫星都是逆行的，但离行星更近时顺行和逆行的卫星是混杂在一起的，而不是像木星的卫星那样整齐地分开。

73

土星有39颗已知的不规则卫星，包括土卫七（图11），一个半径为180千米×133千米×103千米的卫星，它的轨道在两颗规则卫星土卫六和土卫八之间。土卫七在已知的卫星中是独一无二的，因为它的自转是混乱的。它的自转周期是变化的，甚至它的自转轴也会随着它的滚动而改变。它的密度只有固态冰的一半，内部可能是多孔的碎石。它的反照率很低，这表明它的表面有黑色粒子的尘埃，这是这一部分的土星系统的典型特征。

土卫九（图11）的半径为109千米×109千米×102千米，是土星逆行的不规则卫星中最大和最近的卫星。土卫九的轨道位于土星半径548倍那么远的位置。这实在太远了，“卡西尼号”在进入轨道后无法访问，所以“卡西尼号”在接近土星的时候安排好了时间，在2 000千米的距离上近距离掠过土卫九，这使其拍摄到的图片成为所有同类中图像最好的例子。“卡西尼号”发现了一个布满环形山的表面，并探测到了水冰、二氧化碳冰和黏土矿物。土卫九的反照率极低，只有0.06%，这可能是因为甲烷冰中的一些氢被剥离了（长时间暴露在太阳紫外线辐射下可以做到这一点），使得碳原子连接在一起，形成一种黑色的焦油状黏稠物。土卫九是一颗受捕获的半人马型小行星的最可能候选——半人马型小行星是一种多在土星轨道之外发现的冰冻小行星。2009年，红外望远镜的观测结果显示，土卫九的轨道处在一个弥漫但非常宽（比土星本身厚20倍）的尘埃带中，人们认为这是由于微陨石的撞击而从土卫九的表面落下的物质形成的。

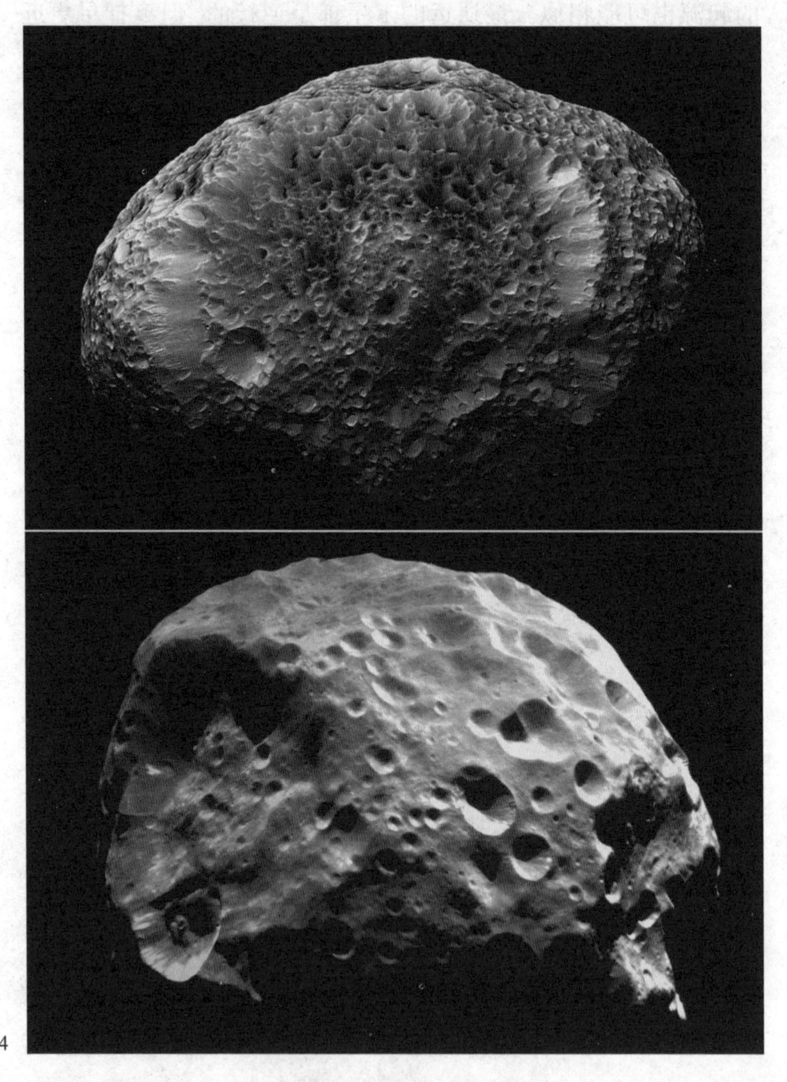

图11　土卫七（上图）和土卫九（下图）的尺寸相似

土星有两组顺行轨道不规则卫星。其中一组（因纽特群）的名字以因纽特神话的名字命名，如土卫二十九（西阿尔那克）；另一组（高卢群）以高卢神话的名字命名，如土卫二十六（阿尔比俄里克斯）。每一组都可能是被碰撞摧毁的更大的卫星的残骸。除了土卫九，土星逆行的不规则卫星（诺尔斯群）都由挪威神话的名字命名，这样的分组也许代表它们是同一颗被捕获的小行星碎片。

即使用最先进的现代望远镜，也很难观测清楚天王星和海王星的不规则卫星。1997年至2003年间，人们发现了九颗天王星的不规则卫星。它们的轨道都是逆行的，除了天卫二十三，它的轨道是所有行星的卫星中最偏心的。最大的是天卫十七，半径约为75千米，而已知的最小的卫星半径约为10千米。没有轨道接近的卫星群，每一个都可能是单独捕获的天体。

目前海王星只有六颗已知的不规则卫星：三颗顺行，三颗逆行。其中最大的是海卫二，半径为170千米。它在1949年被发现，其他的在2002年至2003年被发现。最外围的海卫十（半径为20千米）和海卫十三（半径为30千米）位于偏心逆行轨道上，平均距离为海王星半径的1 885倍和1 954倍。这样的距离十分遥远（它们绕行星运行一周需要9 000多天），但这些轨道是稳定的，因为海王星离太阳更远，所以它的希尔球比木星更大。

海卫二可能是一颗规则卫星被灾难性地摧毁后（也许发生在海卫一捕获事件中）的巨大残存残骸。它的表面有水冰，由于碳等深色剂的作用，它的反照率很低，这一点与天王星的一些规则卫星很相似。有人提出，另一颗不规则卫星——海卫九，可能是来自同一天体的较小碎片，但海王星的其他不规则卫星很可能是单独捕获的天体。

内卫星

内卫星可以非常小，正如它们的名字所暗示的那样，它们靠近其行星导致了眩光，使得它们比不规则卫星更难被望远镜发现。法国天文学家爱德华·洛希（1820—1883）提出了一个很好的理由来解释为什么大的卫星不能在它们的行星附近被发现。他计算出，每颗行星在一定距离时对卫星近地侧和远地侧的潮汐拉力之差将超过卫星自身的重力。这个距离通常被称为“洛希极限”，此时流体或松散的固体会被撕开，尽管固体的内部强度允许它在解体前继续靠近。

大多数内卫星的轨道都在其行星的洛希极限内运行，它们很可能是被潮汐力撕裂的较大卫星的碎片。一些距离更远、体积更大的卫星可能是由规则卫星受碰撞而产生的。

目前已知的木星内卫星只有四颗，其中最大的木卫五在被“旅行者号”拍下之前就被发现了，其中“伽利略号”拍摄的图像最好，可在图12中看到。

土星在土卫一的轨道内有八颗已知的内卫星，在土卫一和土卫二之间还有三颗（它们的半径只有1千米左右）。在这三颗卫星中，只有土卫三十二被“卡西尼号”近距离观测过，而且它有着令人惊讶的光滑的卵形形状。图12中包含了“卡西尼号”拍摄的土星常规内卫星的最佳图像，大多数图像比“伽利略号”在木星提供的图像的细节要好得多。土星最近的内卫星，土卫十八和土卫十五，在赤道周围有脊状突起，这使它们形成了奇怪的飞碟形状。它们的轨道位于土星环系统的间隙，而赤道脊很可能是被环物质扫过形成的。

土卫十和土卫十一是一对独特的天体：其中一个的轨道只比另一个长几千米。当轨道靠内运行更快的卫星赶上另一颗卫星时（这种情况大约每4年发生一次），它们之间的相互引力作用会导致它们交换轨道。之前速度较快的那颗现在位于更远的轨道上，所以它会减速，而之前速度较慢的那颗会加速，直到4年后追上它的同伴，然后重复这个循环。

天王星有十三颗已知的内卫星，其中只有天卫十五（图12）被“旅行者2号”发现并观测了细节。它们都是暗的（低反照率）天体，可能是甲烷受到辐射破坏导致的。它们大多数由“旅行者2号”发现，但其中三颗（包括最小的天卫二十七和天卫二十六，它们的半径只有5千米左右）是最近通过望远镜发现的。这十三颗行星的轨道范围都很窄，处于天王星半径的1.95倍到3.82倍之间。模拟表明，它们会干扰彼此的轨道，因此在未来1亿年内的某个时候很可能会发生碰撞。

77

“旅行者2号”拍摄的图像显示了海王星的六颗内卫星，哈勃太空望远镜拍摄到了第七颗（最小的一颗，半径约为18千米）。与天王星上的同类卫星一样，它们的反照率很低，似乎是被辐射破坏的甲烷使水冰变黑了。其中，海卫八（图12）是太阳系最大的内卫星，其半径为220千米×208千米×202千米。它的形状与流体静力平衡相距甚远，很可能是由于碰撞造成的。海王星最内部的内卫星，海卫三，在洛希极限内运行。它的最终命运可能是螺旋进入海王星的大气层或被潮汐撕裂，形成一个新的环。

78

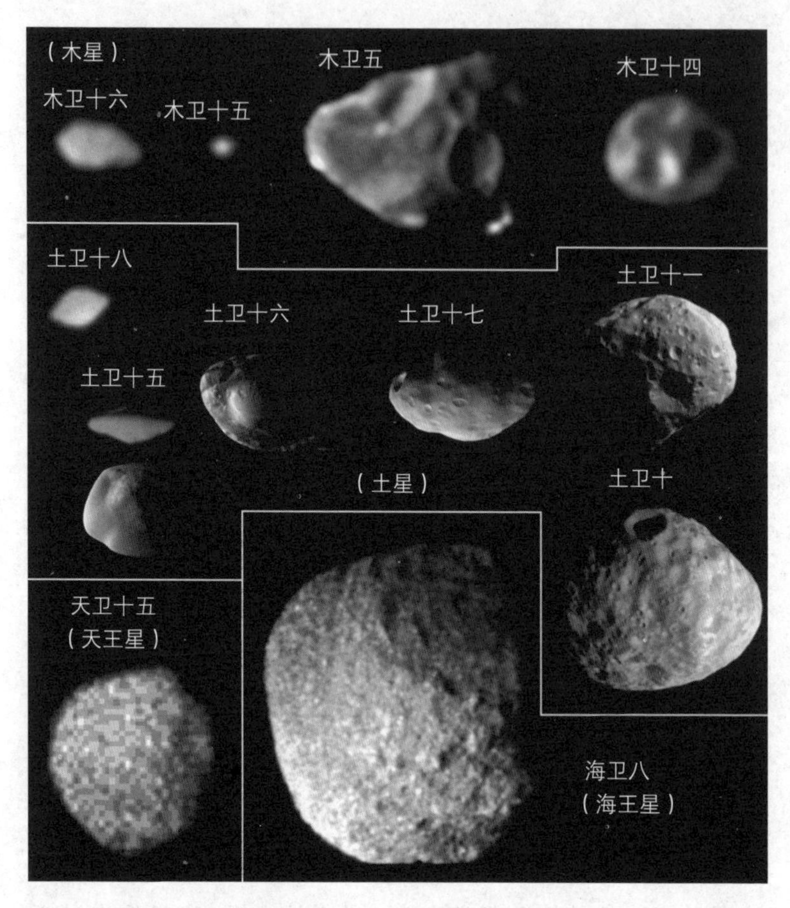

图12　这是一组成像最好的内卫星，以近似正确的相对尺度显示（土卫十七从一端到另一端为100千米）。对于木星的卫星（“伽利略号”图像）和土星的卫星（“卡西尼号”图像），小卫星的排列为越左边的轨道越靠内。土卫十五的两个视图包括：赤道（上图）和南半球（下图）。天卫十五和海卫八的图像来自“旅行者2号”

土星环和牧羊犬卫星

土星以其壮观的光环系统而闻名。伽利略第一次发现了它，但他没能看清楚。惠更斯（在1655年）是第一个正确解释他所看到的东西的人。其他巨行星也有光环，尽管它们的质量要小得多，所以远没有那么壮观。

即使是土星的壮观光环的质量也比它最小的规则卫星土卫一要少。它们可能非常古老，主要来自离年轻的行星太近而没有汇聚成卫星的物质，或是当偏离洛希极限时被撕裂的卫星残骸。很容易看到的环的范围（你可以用小型望远镜自行观测到）由国际天文学联合会指定的C、B和A环组成，它们只有100米厚，从距行星中心的7.5万千米延伸到13.7千米，即与中心的距离从土星半径的1.24倍到2.27倍，也就是距离土星最外层0.24倍半径到1.27倍半径。光谱学表明，这些环大部分由水冰组成，由于辐射破坏或灰尘污染，颜色变暗，大部分地方变红。当光环进入土星的阴影时，相对冷却速率相对较慢，地面雷达能够记录环带反射的信号，还有环带对“旅行者1号”的广播传输造成的影响，它们都表明光环是由从约1厘米至5米的物块组成的。每一块这样的物体都位于环绕行星的轨道上。尽管并没有规定一颗卫星的尺寸下限是多少，但把它们中的每一颗都看作卫星是不合理的。

在主环的两侧出现了次要的环。D环（6.69万千米至7.45万千米）位于C环和行星之间，而其他环位于A环之外。土星环的内部结构非常复杂，最明显的是在距离土星较远的地方，环内的物质变得更加分散，甚至完全消失。最宽的“缝隙”有4 800千米宽，将A环和B环分开。它是由乔凡尼·卡西尼于1675年发现的，被称为“卡西尼环缝”。在它的内缘上，任何组成环的粒子的轨道周期都是土卫一的两倍，这也很容易解释。但缝隙内的精细结构，即粒子集中或分散区域的半径，目前还不完全清楚。

环上的许多小间隙无法用简单的轨道共振解释，但有些确认了间隙被内卫星扫过，这些小卫星的轨道实际上是沿着间隙内移动的——例如，土卫十八和土卫三十五在A环上扫过不同的间隙（图13）。

“卡西尼号”在2013年拍摄的图像显示，在A环的外缘一个长1 200千米、宽10千米的物质增加区域，物质可能集中在一个直径约为1千米的小质量物体周围。这可能是一个正在形成的

新的小卫星，但更可能是一个将会消散的暂时扰动。

F环只有30千米至500千米宽，位于A环之外。这是由于内卫星土卫十六（图13）和内卫星土卫十七（图13）分别在它的内缘和外缘运行，位于二者轨道之间的天体所形成的动态结构。像这样的内卫星，其与环带的密切联系有助于确定卫星的形状，通常被称为“牧羊犬卫星”。

在F环之外，有各种各样的纤细的环和环弧，据信是由微陨石的轰击从最外层的内卫星——土卫五十三、土卫三十二、土卫四十九和土卫三十三上撞击下来的尘埃，通过与土卫一的共振，它们被限制或集中成环弧。在更远的地方，即18万千米到48万千米处，弥散的E环由直径小于1微米的粒子组成，由水冰和从土卫二喷发出来的其他物质的痕迹组成。环绕土卫九轨道的弥漫尘埃带有时被称为“额外的环”，但这个词实在是过誉了。2008年就有一种说法认为，土星的规则卫星土卫五被它自己的弥漫环系统所环绕，但现在这种说法已经不成立了。事实上，没有已知的卫星有环，就像没有卫星有卫星一样。

木星的光环是由“旅行者1号”发现的。主环只有6 500千米宽，30千米至300千米厚。最内层的小卫星木卫十六的轨道

图13　土星环中的卫星。上方：土卫十六，长176千米，从右到左移动，对土星F环造成周期性扰动。中左：土卫十八，34千米长，在恩克环缝轨道上运行，由于切线入射的阳光投下了长长的阴影。中右：土卫三十五，8千米长，在基勒环缝中运行，在A环的内缘产生了波动。底部：天王星的光环上有两颗卫星，天卫六（左）和天卫七（右）。它们的半径约为20千米，在这张长时间曝光的照片中，它们被人为拉长了。图像上有很多噪点，最外层环两侧的暗带是人工处理的

占据了这个环中的一个缺口，而环的外缘则由木卫十五的轨道引领。环内物质是一种未知的红色物质，由微米大小的灰尘和更大的物块组成，总质量不超过木卫十五的5倍，也可能更少。辐射压力和与木星磁场的相互作用将在不到1 000年的时间里驱散尘埃，所以除非它非常年轻，否则它必须通过较大物块之间的碰撞，或通过木卫十六和木卫十五受到的微陨石轰击来补充。

2007年，美国宇航局的“新视野号”的任务是在飞掠土星环时，在环内寻找未知的内卫星，但未能成功发现，所以确实不太可能存在半径上超过0.3千米的小卫星。然而，它确实发现了七个物质团块，占据了绵延1 000千米到3 000千米的环弧。

在主环的内部，木星有一个12 500千米厚的“光环”，它似乎是由主环螺旋向木星内部移动的尘埃构成的。还有两个非常微弱的“薄纱”环，它们是从最外层的内卫星木卫五和木卫十四的轨道向外延伸的尘埃，可能是由微陨石轰击小卫星的表面喷出的物质补充形成的。

天王星光环于1977年被天文望远镜观测者发现，他们注意到，当一颗星星被原本看不见的光环穿过时，它会不断变暗。我们对天王星环的大部分了解来自“旅行者2号”和哈勃太空望远镜。目前已知天王星有13个环，总质量超过了木星的光环，但比土星的光环小得多。有几个环带很窄，主要是低反照率的红色物质组成的卵石大小的物块，据信是由内卫星的碎片产生的。在所有这些狭窄的环带中，最外层和最亮的环（ε环）由小卫星天卫六和天卫七引领（图13）。它与行星中心的平均距离为51 150千米，但由于它的形状是偏心的，因此围绕它一圈，离中心的距离变化了800千米。它离行星最近的地方是20千米宽，离行星最远的地方是100千米宽。内环的约束更紧密，但没有已知的牧羊犬卫星，这促使人们认为牧羊犬卫星是在大约6亿年前由碎片形成的。

天王星的其他光环都由尘埃组成。在窄环内部的有四个，据信是螺旋向行星运动的短命尘埃，在窄环外部有两个——最外层的尘埃散布在最外层的内卫星天卫二十六的轨道上，其可能是尘埃的来源。

83

海王星有五个环，和天王星的环一样非常暗。它们大约一半是灰尘，一半是更大的物块。最外层环上的物质集中成几个环弧，人们试图将这解释为来自内卫星和牧羊犬卫星海卫六的42∶43共振，但这种尝试并没有成功。很明显，关于环以及它们与卫星的关系，我们还不太了解。

84

第五章

近距离观察规则卫星

25年前，我写了一本以“旅行者号”为基础的书，将巨行星的规则卫星描述为“完全只属于它们的独立世界”。这句话的真实性现在更加明显了。它们中的每一颗都很吸引人，其中两颗甚至被广泛认为是比火星更适合外星生命的候选者。在这里我将讨论一些我最喜欢的卫星。

木卫一

由于潮汐加热，木卫一是太阳系中火山活动最活跃的天体，其火山活动活跃度甚至超过了地球。“旅行者号”探测器发现了九个火山喷发羽状物，其中最大的一个高300千米，宽1 000多千米。“旅行者号”探测到来自羽状物源的强烈红外辐射，证明了木卫一上的高温。在“旅行者号”和“伽利略号”之间的16年期间，以及2007年“新视野号”飞掠的前后一段时间里，地球上越来越精密的红外望远镜记录下了这些以及其他许多木卫一上的“热点”。

木卫一上活火山的发现促使国际天文学联合会修改了木卫一上的特色命名规则，改为神话中的铁匠与火、雷、太阳、火山神和英雄的名字。例如，对于图14中的喷发的活火山，选择了“普罗米修斯”来命名，他是一位希腊神明，偷走了火并交给人类；特瓦史塔火山口是以梵文史诗中的太阳神来命名的；产灵火山以一个日本的火山神来命名。现在已经记录了大约100个火山爆发的地点，其中一些（如普罗米修斯）是持续性的，其他一些（如特瓦史塔和产灵）已经爆发过几次。 85

木卫一表面没有任何可见的环形山，因为它不断地被火山活动重塑表面，陨击坑总被迅速掩盖。这种表面重塑是由火山喷发羽状物的沉降物和熔岩流的冲刷/淹没共同作用的结果。木卫一表面占主体且浓重的黄色和橙色最初被认为表明了许多或所有的熔岩流成分中有硫，但对活跃火山口的红外测量显示，它们内部的温度对于硫来说太高了。炽热的物质一定是熔融的岩石，可能是低黏度的岩浆，硅含量低，镁含量高。

低反照率熔岩流具有可识别的细长的叶状特征，可长达300千米，经常出现在火山口状的地貌中，其圆周既不光滑，也不够圆，不足以被认为是环形山。当熔岩流停止移动后，它开始变得不可见，因为它被硫黄和二氧化硫霜（这塑造了木卫一的颜色）所覆盖，这些霜都是从地面上喷出的，并作为火山喷发羽状物的尘埃沉积下来。

有些火山喷口没有岩浆流出，却是强烈的红外辐射源，显然在内部含有“熔岩湖”，岩浆不断从中涌出，覆盖地壳，再下沉，最后被下面的新岩浆所取代。 86

其他的火山喷口或裂缝是木卫一最大的羽状物的来源。它

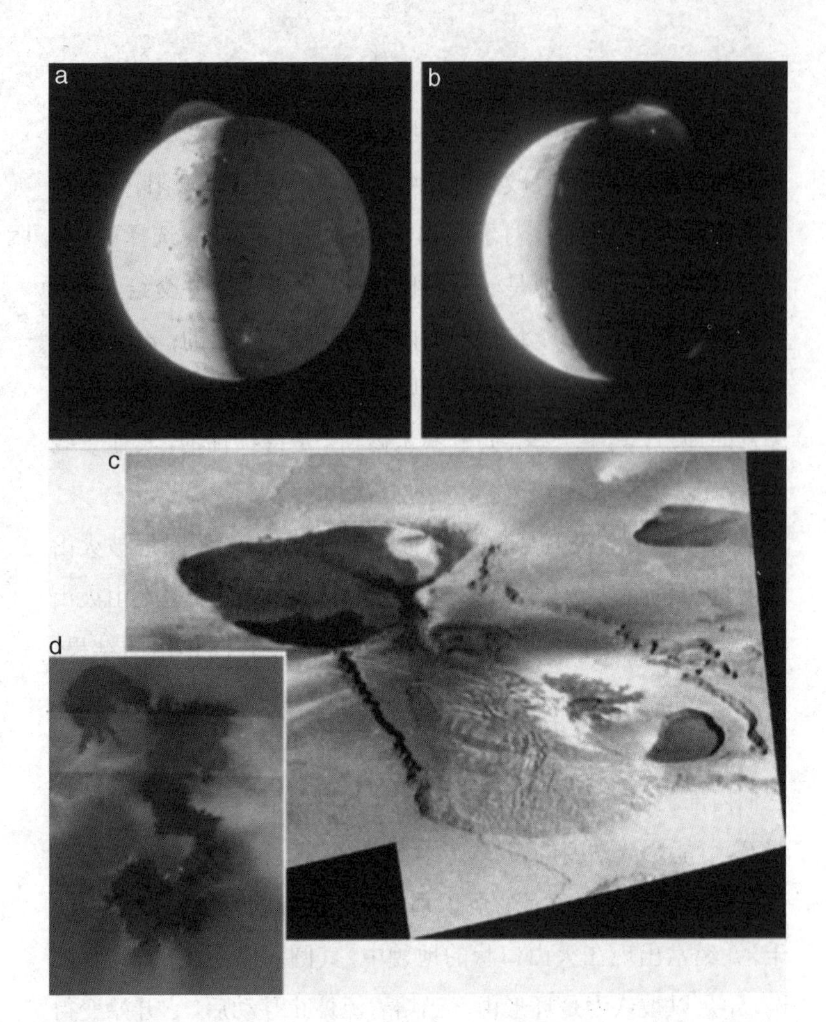

图14 (a)及(b)为"新视野号"于2007年2月28日及3月1日拍摄的木卫一。来自特瓦史塔火山的290千米高的喷发羽状物在北部非常突出。2月28日的这张照片还显示了一个长达60千米的羽状物,来自9点钟位置边缘的普罗米修斯火山,而顶部的第三个羽状物,来自产灵火山,其上升得足够高,可以捕捉到南方的阳光。木卫一的夜侧被木星反射的光线照亮,但白天的一面被人为过度曝光,以增加羽状物的能见度。在3月1日的图像中,可以看到特瓦史塔火山的喷口在黑暗中发光,在4点钟方位的黑暗边缘上可以看到产灵火山羽状物。(c)为"伽利略号"在2001年10月16日拍摄的300千米宽的特瓦史塔火山图像拼接图。阴影的变化令人困惑,因为一天中不同时间的图像被组合在一起。(d)为由"伽利略号"拍摄的90千米宽的拼接图像,图中显示岩浆流从火山口喷发出来(左上角)。普罗米修斯火山羽状物起源于主要熔岩流的前端附近,也就是图像的底部

们可能经由岩浆中的气泡膨胀来提供动力，当气泡穿过地壳上升时，气体膨胀的力度足以粉碎岩浆碎片并把它推向天空。在地球上，这一过程会吸引周围的空气，并发展成一个对流柱，但木卫一实际上没有大气，所以粒子沿着抛物线弧线上升，然后回到地面。图14（a）和14（b）中的特瓦史塔火山的羽状物很好地证明了这一点，在图14（a）和图14（b）中，喷发速度约为每秒1千米。推动火山喷发的气体是硫黄和二氧化硫的混合物。这两种物质在它们的轨道顶部附近凝结成“霜”颗粒，并与岩浆碎片一起落到地面，但在这种情况下，岩浆占主导地位，因此沉积物的反照率很低。

普罗米修斯火山羽状物和产灵火山羽状物是另一种类型。在这种类型中，火山羽状物并非起源于火山口，而是起源于熔岩流前进的前沿。在这种情况下，炽热的熔岩蒸发了覆盖表面的二氧化硫霜。正是这一点形成了羽状物，当膨胀的气体凝结成霜状颗粒时，羽状物变得可见。这类羽状物主要由二氧化硫构成，有时伴随着少量岩浆，有时则根本没有岩浆。它们坠落的地方会形成高反照率的沉积物。在图14（a）中可以看到许多这样的白斑，特别是在被木星照亮的夜侧区域。

木卫一的喷发平均每秒释放1吨左右的氧和硫，它们在与木星磁场（包围了木卫一）的相互作用下被电离，并形成一个甜甜圈形状的“等离子体环”，在木卫一轨道上环绕着木星。

88

木卫二

木卫二与木卫一看起来几乎没有什么不同（图15），但如果我们能剥去约100千米厚的冰和水，我们可能会发现一个在其之

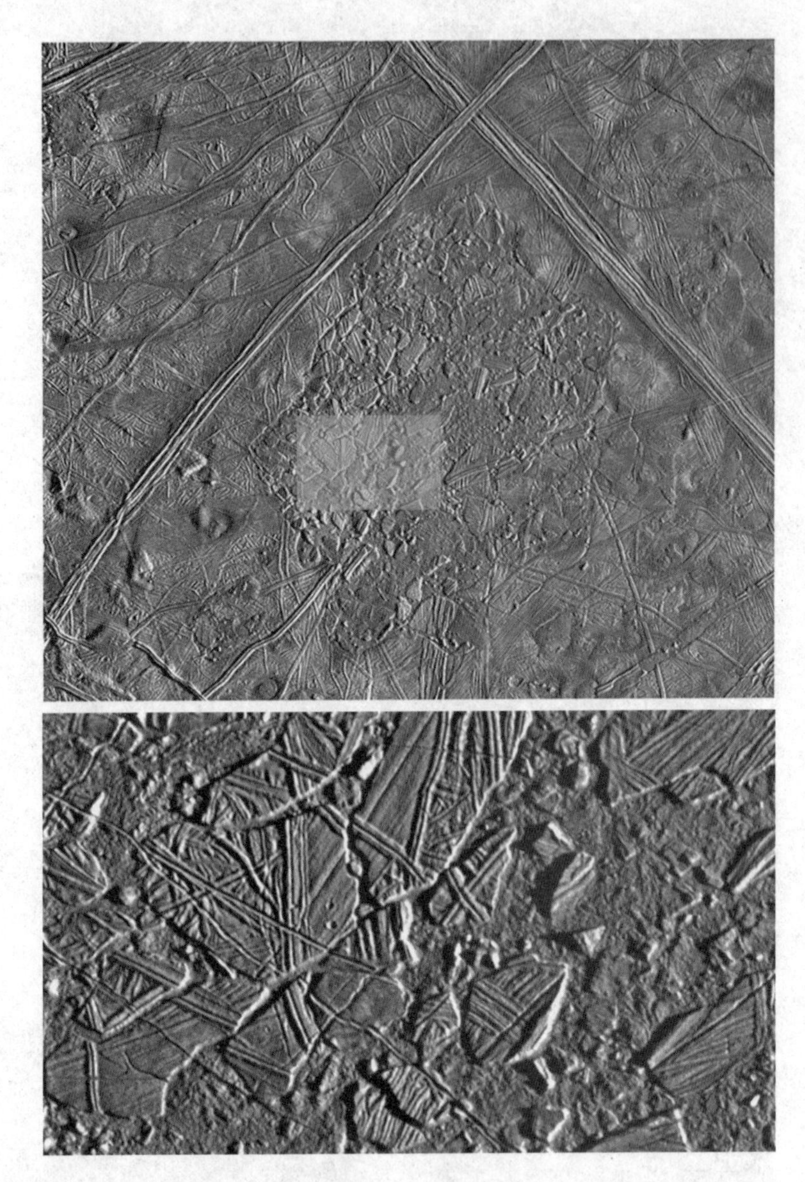

图15　由“伽利略号”拍摄的木卫二。上面的图像是一个300千米宽的区域，显示了“一团线球”的地形，中心区域除外，它已被分解形成“破碎地貌”。透明矩形中的区域以更高的分辨率显示如下图，显示了漂浮物和基质。下图中两个最大的环形山直径不到200米，就在中间偏左的位置

下不那么活跃的木卫一。木卫二的表面是高反照率的冰，它太过年轻，以至还没有积累几个陨击坑。表面上有一大部分都是大量山脊和沟脊，它的外观被比作“一团线球”。大多数山脊似乎都位于裂缝打开后又闭合的地方，裂缝中挤压出一些雪泥在表面形成山脊。这个打开和关闭的过程可能是一个潮汐过程，因此木卫二每绕木星运行85小时，这个过程就会重复一次，在裂缝的有效寿命内形成一个100米高的重新冻结的冰脊，然后它会永久闭合，在其他地方形成一个新的裂缝。一些特别大的山脊被变色的冰环绕着，显示出在结冰的水里有杂质（可能是硫酸镁和其他盐）。

线球地形在不同的区域被破坏，主要是被分解形成一种被称为“破碎地貌”的地形。在这里，你可以看到带有线球纹理的漂浮物，它们被一个有着复杂纹理的低洼表面分开。现在，大多数科学家都认为，这些漂浮物之间的“基质”是水体重新冻结的表面，当表面冰变薄时，水体暂时暴露在冰面上，使其分裂成漂流分开的浮冰。在一些混沌区域，人们可以看到漂浮物曾经是如何组合在一起的，但在另一些地区，整个漂浮物要么沉没了，要么融化了。在较早的混乱区域，一些新的山脊和沟脊穿过类似漂浮物和基质的地形，这表明破碎地貌最终变得无法识别，并被新一代的线球地形覆盖。

即使是非常咸的水，一旦暴露在木卫二零下160摄氏度的表面温度下，也会迅速冻结，但是我们可以一块想象数千米厚的漂流在数十米厚的泥浆海面上的浮冰，它漂流了好几个星期，之后冰泥浆变得太厚、太硬，无法继续漂流。根据漂流物表面和基质表面之间典型的数百米高度差，并考虑到漂浮物和盐水之间可能的浮力差，计算表明，当冰盖断裂成漂浮物时，其厚度在五百米到几千米之间。

专家们还没有达成一致意见的是，每个破碎地貌形成时暴露出来的水体是木卫二地下的全球海洋的顶部（图9），还是仅仅是木卫二冰壳内一个孤立的液态水透镜体的顶部。我们无法得知冰壳的厚度（最好的方法是用木卫二的轨道飞行器来绘制潮汐隆起的高度，并配备穿冰雷达）。冰壳可能有几十千米厚，在这种情况下，融化过程将从底部向上使外壳变薄，直到它足够薄而破裂，这需要持续的且能够融化厚度达几十千米的冰的局部热量输入，然而，如果外壳只有几千米厚，融化就会容易得多。推动这一过程的热量可能是木卫二的岩石内部产生的潮汐加热，可能是由海底火山喷发传递到地面的。

“木卫二厚冰模型”的支持者更倾向于在冰中形成一个暂时的液态水透镜体，通过深层冰中固态对流的热量向上输送而在冰中融化。这可能是由于冰内的潮汐加热，也可能是由于海底火山爆发向海洋输送的热量。

尽管木卫二基本上没有大气层，但它是一个比火星更有希望找到外星生命的地方，虽然不那么方便。地球上的生命据信来自海底的热液喷口，而木卫二岩石地壳的潮汐加热提供热量，防止了地下海洋冻结，这也会形成大量的热液喷口，水在被岩石向内吸收并加热后排出。

在地球的深海中，有些生态系统是靠热液喷口释放的化学能生存的。它们完全不依赖于植物用来新陈代谢的阳光。在木卫二海底的喷口周围存在类似生物是完全可能的，它们可能是食物链底部的“化学合成”微生物，还有以它们为食的更大的、可能是多细胞的捕食者。

如果生命起源于木卫二上的深海喷口，那么它可能随后适应了在其他我们更容易探索的环境中定居。最好的环境也许是潮汐裂缝，每次打开时，水都会被吸进来。任何到达上层几米的浮游生物都能找到足够的阳光进行光合作用，只要它们停留在海面以下至少几厘米的地方，水就能保护它们免受地表可能遭受的严重辐射剂量的影响。每当裂缝闭合时，一些不幸的生物就会随着雪泥被挤压出来，所以最容易找到生命化石的地方就是被埋在构成表面山脊的冰冻雪泥中的地方。

木卫三和木卫四

木卫三也可能在岩石层的上层有海洋，这是供给任何生命赖以为生的化学能的重要因素。如图9所示，木卫三有可能被几个连续的固体和液体壳层所覆盖。然而，无论确切情况如何，即使是最顶层的海洋的顶部也在远离表面的地底深处。木卫三表面没有可见的破碎地貌的痕迹，所以可能它表面的冰壳还没有薄到足以融化。然而，也有过多次类似的平行沟脊群形成的事件。最年轻的样本横切穿过古老地形，这里比其他地区有着更高的反照率（图16）。这可能是在木卫三上相当于线球地形的地方。但大多数沟脊只是裂缝，没有在木卫二上看到的山脊。

木卫三与木卫二的一个重要区别是，木卫三的表面破裂似乎很久以前就停止了。图16显示了最年轻的高反照率地带上的许多直径为2千米以下的环形山。更古老、更黑暗的地表上也有这样的环形山，但图片中也有直径高达10千米大小的环

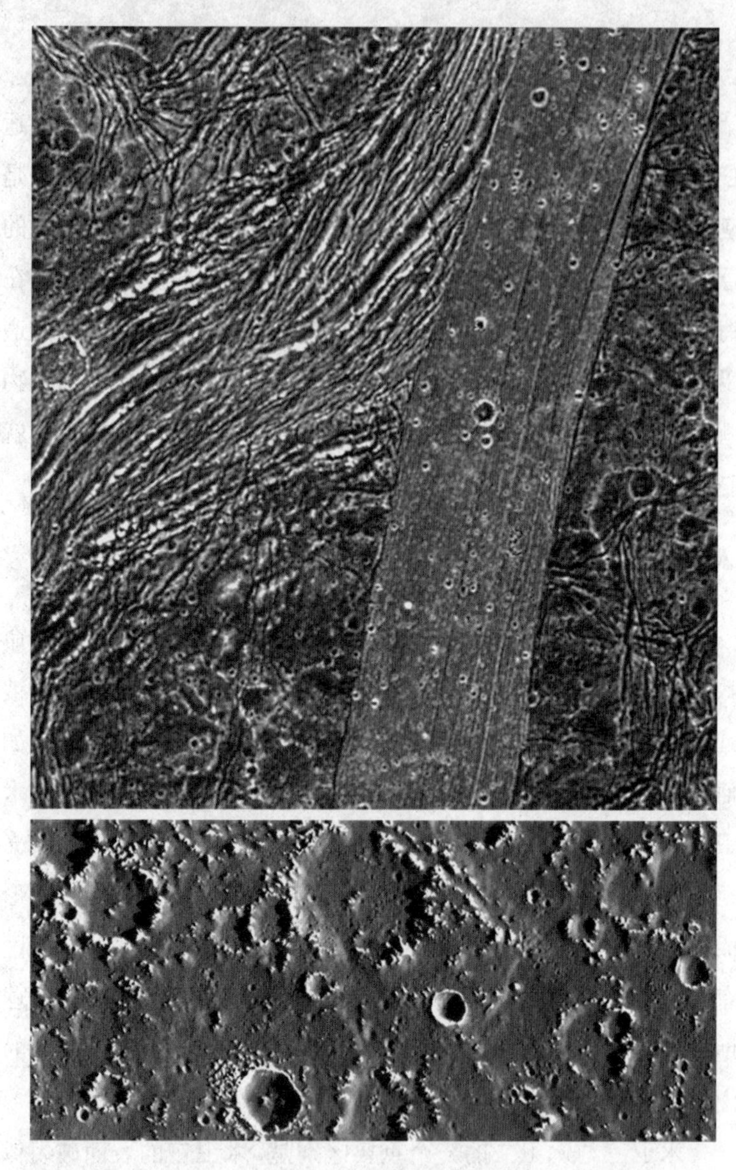

图16　木卫三（上）和木卫四（下）上70千米宽的代表性区域，由“伽利略号”拍摄。木卫三图像上最年轻的地形是南北贯穿的高反照率地带，但即使是这样的地形也一定非常古老，因为造成它上面的许多陨击坑需要很长的时间。请注意，在木卫三的图像中，太阳高悬在西边，而在木卫四的图像中，太阳在东边更低处

形山，这些环形山被许多裂缝切割。在图16所示的区域之外，木卫三有大小达200千米的环形山，其中一些是位于高反照率地带上的。

虽然木卫三表面的环形山密度是令人信服的证据，表明该表面是古老的，但遗憾的是，我们不能用与月球环形山相同的时间尺度算法来得出其绝对年龄。这是因为如果你绘制出木卫三环形山出现率与环形山大小的分布图，它将看起来与月球上的同类图非常不同，这表明木卫三受到了不同种类的撞击。此外，没有理由认为木星被撞击的历史和频率与地球一样，因为地球到太阳的距离比木星到太阳的距离近5倍。

然而，我们可以得出这样的结论：木卫三和木卫二环形山密度的巨大差异提供了明确的证据，表明木卫三表面的平均年龄比木卫二要大得多。木卫三的地表年龄很可能超过30亿年，而木卫二上现存任何地表的年龄都不可能超过5 000万年。

木卫四表面的陨击坑比木卫三的陨击坑还要多，即使木卫四的表面确实有裂缝，这些裂缝也被陨击坑抹去了。图16显示了木卫四的一个区域，其规模与木卫三的照片中的区域相同，存在半径达到15千米宽的环形山，但在其他地方还有很多更大的环形山。事实上，木卫四拥有太阳系中最大的环形山，即一个直径3 800千米名为瓦哈拉的多重环形盆地。虽然瓦哈拉盆地的环形山幸存了下来，但它不再保持原来的深度，并且变得扁平，因为木卫四的冰层无法维持如此深的盆地，因此向上流动以修复撞击的破坏。木卫四内部的海洋（图9）是由其感应磁场推断出来的。它对木卫四的表面没有明显的影响，所以其位置一定比木卫二的海洋更深。

土卫六

尽管土星有七颗规则卫星，但土卫六是唯一一颗在大小上能与木星的伽利略卫星相匹敌的卫星。它的大小只比木卫三稍微小一点，质量也比木卫三小一点（见附录），并且有一个冰构成的外壳。“卡西尼号”的观测显示，在这个外壳约100千米深的地方，有一片地底海洋将它与行星的内部隔开。如果这片海洋是水和氨混合的，那它可能有200千米厚，在它下面将是一层200千米厚的冰Ⅵ，围绕着土卫六的岩石核心。

土卫六的特别之处在于，它在卫星中是独一无二的：它有着稠密的大气层，其表面大气压是地球的约1.5倍。其中大约98%是氮，剩下大部分是甲烷。在离地面50千米到250千米的地方，太阳紫外线将甲烷中的氢原子分解（这个过程被称为光解作用），然后碳原子连接成链，形成烃类烟雾，导致在大多数波长的光线下都看不到星球表面。“旅行者号”未能观测到土卫六的地表，但“卡西尼号”携带了专门用来穿透这片烟雾并绘制地形的成像雷达。此外，“卡西尼号”的光学相机能够使用一种特殊的滤镜获得地表的低分辨率视图，这种滤镜被证明对跟踪地表的季节性变化很有用。我们还从“惠更斯号”探测器（由“卡西尼号”携带到土卫六）上看到了土卫六的地表，它在降落伞的降落过程中和处于地面的过程中发回了烟雾下方的清晰照片。

这揭示了一个怎样的地表啊！如果你不知道土卫六表面的“岩石”实际上是水冰，稀疏蓬松的低空云层是浓缩的甲烷和/或乙烷，你可能会认为你看到的是地球的一部分。土卫六的表面会经历甲烷降雨，这些降雨聚集在小溪和河流中并侵蚀山谷，流入湖泊，其中三个湖泊大到足以被称为海洋，如图17中的克拉肯海。这些海洋和大部分湖泊都位于北部高纬度地区，但也有湖泊位于南极附近。此外土卫六上还发现了干涸的湖床，这显示存在一些季节变化的迹象。湖泊水量变化的证据也来自它们的海岸线特征；例如，丽姬娅海南岸的大部分海岸看起来就像一条典型的被淹没的海岸线，在那里，干燥陆地上由外力刻出的丘陵和山谷都被海平面上升淹没了。还有可能存在泉水形成的池塘和湖泊，流入其中的液体可能来自甲烷雨，这些甲烷雨渗入冰层并渗透到地下（通过在地球上被称为“含水层”的地方），其中一些可以在再次出现在地表之前转化为乙烷或丙烷。

“惠更斯号”着陆器在土卫六赤道以南不远的地方着陆。它在向下的过程中看到了干涸的河谷，并降落在泛滥平原或干涸的湖床上。这片区域布满了由甲烷河的运输而形成的冰卵石，这些冰卵石坐落在深色的基石上，这些基石可能是雨水冲刷出来的碳氢化合物形成的焦油混合物。

在低纬度地区，大片的低反照率沙丘覆盖着冰的“基岩”，单个的沙丘大约有100米高，数百千米长。这些沙丘似乎是由来自土星的潮汐力推动，与向东吹的风平行而形成的。这些沙丘中的“沙粒”可能是被污染的冰粒，或碳氢化合物的颗粒。

尽管土卫六的大气层高度和密度很大，但大型撞击物应该能够以制造陨击坑所需的高速到达土卫六表面，虽然目前只发现了八个陨击坑，直径从29千米到292千米不等。环形山的稀少表明地表的平均年龄还很年轻，但我们不知道那些较陈旧的环形山是否只是被（河流和风的）侵蚀抹去，被沉积物掩埋，或者是否还有其他过程在起作用——其中之一可能是火山活动。

图17　地图显示的土卫六北极附近450千米宽的区域，由“卡西尼号”在2004年至2013年间收集到的不同质量的成像雷达图像汇制而成。图的左边是北方。北纬80度的纬度线和间隔为10度的经度线叠加在一起。大片的黑暗区域是丽姬娅海，其北岸附近深度不足5米，使得雷达波可以从海底反射，但其他地方深度超过100米。可以看到，有好几条河流流入其中

土卫六上有几个候选火山，比如索特拉帕特拉，这是一座1.5千米高的双峰山，拥有火山口且已经出现了明显的熔岩流。这里 97 的熔岩不是熔融的岩石，而是土卫六的冰地壳部分融化后形成的。它可能是一种氨-水混合物（在比纯水温度低得多时处于液态），或者甚至是一种蜡状物质，通过大气降雨导致埋于地下的碳氢化合物的反应而形成。

为了区别于地球、月球和木卫一等传统的涉及熔融岩石的火山活动，人们将喷发物质为任何种类的冰的火山活动称为冰火山活动。在土卫六上，似乎不太可能发生明显的潮汐加热，所以驱动力可能是内部岩石放射性衰变产生的热量。

土卫二

土卫二是土星规则卫星中除了它的内邻土卫一之外最小的一颗。然而，土卫一上布满了环形山，土卫二则是一个更有吸引力的地方。“旅行者号”显示，土卫二的一部分地区是坑坑洼洼的地形，另一部分地区是看起来平滑而毫无特色的地形。“卡西尼号”获得的高分辨率成像显示，那些表面光滑的区域实际上是充满裂缝的，而且曾经在那里的大部分陨击坑已经被几代的交叉裂缝抹去了。

更令人兴奋的是，“卡西尼号”发现，土卫二南极附近的一系列裂缝中爆发出了羽状物（图18），这种裂缝被通俗地称为“虎纹”，因为沿着它们的表面显露出略蓝且更新鲜的冰带。目前已经发现了100多个独立的间歇泉，它们以每秒1千米的速度将微小的水冰晶体喷射到太空中。这些晶体显然是形成土星E环的物质来源。“卡西尼号”在设计时没有预料到这些羽状物，但是 98

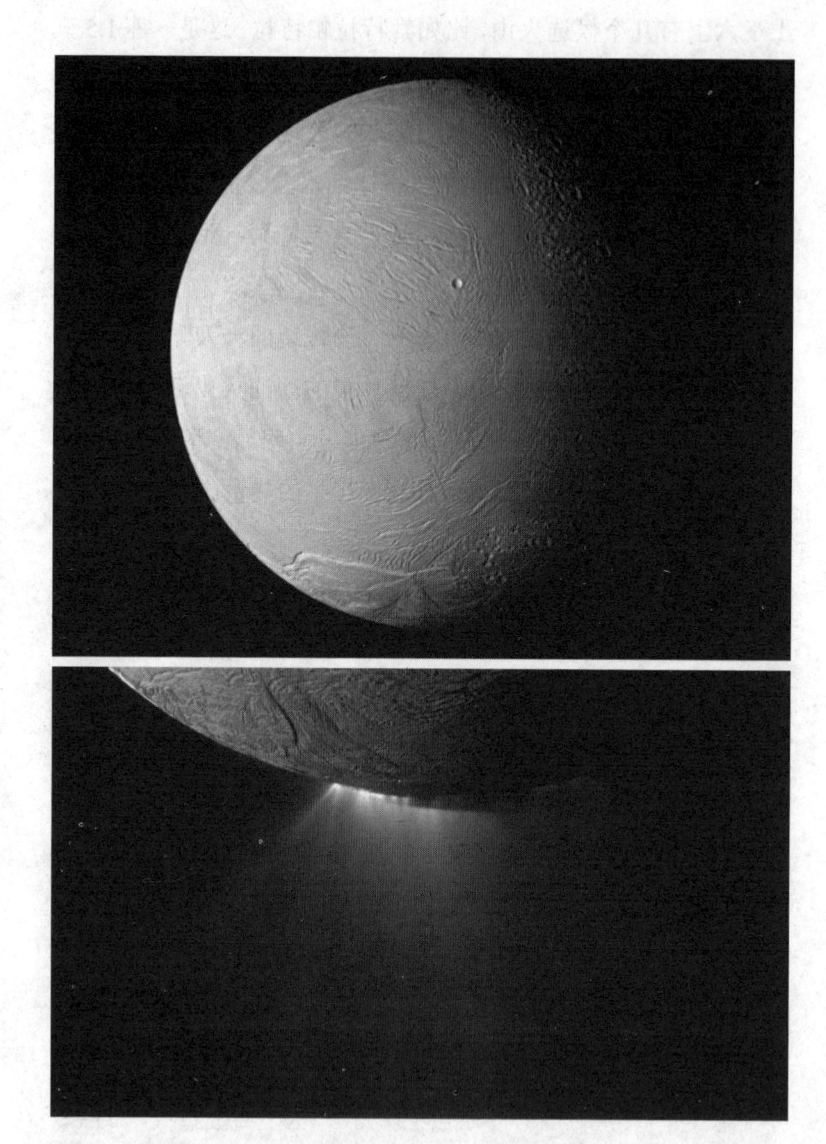

图18　2009年11月21日“卡西尼号”观测到的土卫二（上图）；由“卡西尼号”拍摄于2010年11月30日的土卫二南极地区的照片（下图），显示阳光照射下的羽状物从三个次平行的裂缝中浮现出来

99

任务计划受到了修改，允许“卡西尼号”在离地面200千米到25千米的范围内俯冲穿过这些羽状物。“卡西尼号”的离子和中性质谱仪（主要作为采样土卫六外逸层的离子、原子和与土星磁层相关的分子的工具）表明，羽状物中99%是水，还有微量的甲烷、氨、一氧化碳、二氧化碳和各种简单的有机分子。“卡西尼号”的红外光谱仪显示出，10米宽的羽状物喷射裂缝内的温度至少可达到零下70摄氏度，而该地区的地表温度为零下180摄氏度。土卫二的轨道略呈椭圆形，当土卫二离土星最远的时候，羽状物的强度最大，这时地壳内的压力往往会将虎纹裂缝拉开。

这些喷发物质的来源被认为是土卫二南极下面一大片被潮汐加热而融化的水。潮汐裂缝可能是造成土卫二地表裂缝历史极其复杂的原因。目前的潮汐加热速率是否足以维持目前的地质活动还远不能确定，这可能是先前一段时期中更剧烈的加热活动遗留下来的。

与木卫二不同，土卫二可能没有一个全球性的海洋，液态水洞窟也不太可能存在于岩石中。小范围的水和岩石之间有限的化学相互作用，意味着土卫二上存在微生物生命的可能性小于木卫二。然而，羽状物会将样品送到太空，不需要着陆就可以收集，这意味着未来的飞掠探测器或轨道探测器可以使用专门设计的仪器来搜索有机分子或生物驱动的同位素分馏，而不需要着陆在表面。

在土卫二的羽状物被发现之前，“伽利略号”从1995年到2003年一直在寻找木卫二喷发的羽状物，但没有成功；“卡西尼号”2001年飞越木星时也没有发现任何羽状物。2012年，哈勃太空望远镜探测到木卫二一侧有一个氧原子弥散区，最简单的解释是，那是羽状物中喷出的水的光解作用的产物。然而，2014年的多次探测结果都是空白，如果木卫二确实出现了喷发羽状物，那么很明显，它们一定比目前在土卫二出现的羽状物喷发程度更弱，持续时间更短。

100

不管木卫二和土卫二的液态水地带是否有生命存在，可以想象的是，如果那里有合适的地球微生物，它们就能够在那里生存和繁殖。太空探测器在发射前要进行清洁和消毒，但不可能清除或杀死所有的微生物，有些微生物可以在太空中存活数年。如果“伽利略号”和“卡西尼号”这样的探测器被遗弃在轨道上，有可能有一天它们会撞向木卫二或土卫二，无意中把来自地球且尚存活的微生物带到那里。为了防止这种情况，“伽利略号”在2003年寿命结束时被故意撞向木星，“卡西尼号”也将于2017年在土星遭遇类似的命运。

这不仅仅是保护外星生态系统的道德问题，还应该是任何所谓“行星保护”协议背后的一个重要考虑因素。每一种这样的生态系统都是我们想研究的对象，因此应避免任何飞船上的“偷渡者”会造成的混乱。如果这些生态系统存在，那么重要的是要确定那里的生命是否独立于地球上的生命独立起源，或者太阳系内的生命是否因作为陨石内的意外乘客从一个世界传播到另一个世界等原因，拥有一个共同的起源。

生命在我们的太阳系中是否不止一次地独立起源，这是超越一切的重要问题。除非我们在某处找到独立的生命，否则地球上的生命可能只是一个极其罕见的统计学巧合。但是，如果生命独立地起源于一颗冰冷的卫星，而不依赖于地球上的生命，那么可以肯定的是，生命也已经在银河系的其他宜居星球上诞生了。一项致力于在土卫二或木卫二寻找生命的任务，可能是我们修正“我们在宇宙中是孤独的吗？”这个问题答案的最好机会——从“也许”到“并不”！

101

土卫八

土卫八环绕土星的距离几乎是土卫六的3倍，这使它成为所有巨行星的规则卫星中距离最远的。它的不同寻常之处在于它的轨道与土星赤道的倾角为15.5度，因此它是土星唯一的一颗从表面有时可以清楚地看到土星光环的规则卫星。

当乔凡尼·卡西尼在1671年发现土卫八时，他只有在天空中土卫八位于土星西边时才能看到它。他花了30多年的时间尝试，才隐约瞥见了它在土星东边时发出的微弱光芒。卡西尼正确地解释说，这证明了土卫八在同步自转，而它的前半球比后半球的反照率要低得多。

多亏了“卡西尼号”宇宙飞船，我们现在知道土卫八地表反照率的分布类似于网球表面的模式（图19）。高反照率带（0.5—0.6）通过后半球的中心从一个极点延伸到另一个极点。低反照率区域（0.03—0.05）是一条围绕赤道的宽阔带，其中心在前半球，并被命名为“卡西尼区”，以纪念乔凡尼·卡西尼。土卫八上的所有其他地区都是以中世纪法国史诗《罗兰之歌》中的人物和地点命名的，这首诗讲述了法兰克人和撒拉森人之间的冲突。除“卡西尼区”外，高反照率区一般以法兰克人物命名，低反照率区一般以撒拉森人物命名。

土卫八在很大程度上是一个双色的世界，即使在30米左右的最精细的图像分辨率下也没有灰色的阴影。这两种颜色的区

102

图19 “卡西尼号”看到的土卫八。上侧为前半球，下侧为后半球。亮面和暗面之间的反照率对比实际上比这里看到的更加极端

103

分，可能是由土卫八的前半球扫过来自土卫九或较小不规则卫星的撞击产生的尘埃造成的。但是，主要暗带的尖锐边缘和两个暗带之间的过渡地带的小斑块区域表明，这不是故事的全貌。这些黑色物质可能是一种不到半米厚的富含碳的“滞后沉积物”，是由冰内部的杂质积聚起来的。当表面的冰非常缓慢地升华到太空中（这意味着直接从固体变成气体而不融化）时，杂质就留存了下来。土卫八的缓慢自转（用时79天）意味着它的表面在正对太阳时相比其他自转速度更快的土星规则卫星有着更长的升温时间。

较暗的表面比反照率更高的表面吸收更多的阳光和热量，因此，一旦反照率差异建立起来，它将随着时间的推移变得更强，直到无法挥发的黑色滞后沉积物厚到冰不能再从它下面升华。目前，卡西尼地区赤道正午的温度约为零下144摄氏度，比高反照率地形的温度高约16摄氏度。

土卫八有一个古老而布满环形山的表面。黑色物质位于“温暖”的环形山底面上，而寒冷的、面向两极的环形山内壁为高反照率和低反照率之间的过渡区，并保持明亮。土卫八的其他特点在于它的形状，它的极半径比赤道半径少34千米。考虑到土卫八的低密度，这种极圈扁平化的程度只相当于一个自转周期仅为10小时的天体的平衡形状，因此这个形状似乎在潮汐阻力迫使其自转与轨道周期同步之前就固定了。

此外，土卫八有一个狭窄的赤道山脊，高约13千米，山峰高度可达20千米。在卡西尼地区可以找到1 300千米长的山脊，如图19所示，但它在高反照率地带只有孤立的10千米高山峰出现。它一定很古老，因为它上面覆盖着无数陨击坑，它的起源则是个谜。这可能是土卫八以前的快速自转留下的特征。或者，它可能与土卫十五和土卫十八行星上的赤道脊有一些共同之处，但在这种情况下，它不可能是由土星光环中的物质聚集而来，而必须是由其自身早已消失的光环中的物质聚集到表面而形成。

天卫五

在土星之外，我们来到了只有“旅行者2号”在飞行中访问过的巨行星卫星。任务计划人员在其飞过天王星时遇到了一个特别的问题：天王星的自转轴和规则卫星的轨道与它本身绕太阳的轨道之间，有98度的倾斜角。

从技术上讲，这意味着天王星的自转是逆行的。当卫星的运行方向与行星的自转方向相同时，它们被视为顺行。然而更重要的一点是，1986年1月，当“旅行者2号”飞过天王星时，天王星的南极或多或少正对着飞船到达的方向，因此，这些卫星的轨道在飞船上看就像围绕靶子的环一样。所以，“旅行者2号”不可能像观测木星和土星那样，依次穿过每一颗卫星的轨道，并对交会的时间进行安排，使尽可能多的卫星运行到探测器经过的地方附近。相反，为了最大化对天王星的科学研究，我们所能做的就是近距离飞近这颗目前已知的最内侧卫星，并尽可能地利用其他卫星的远距离图像。对它们的科学研究也受到了限制，因为每个天体的北部都处于长期的季节性黑暗中，根本无法成像。

已知天王星最内侧的卫星是天卫五。它的半径只有234千米，而且现在还缺乏轨道共振，所以它被认为是一个相当沉闷、坑坑洼洼的物体。然而，事实证明它非常复杂，而且非常有趣（图20）。大约有一半的受阳光照射的半球具有相同的反照率，而且环形山密布，较老的陨击坑有着柔和的轮廓，就像被一层土褐色的物质覆盖了一样，而较年轻的陨击坑则有锐利、清晰的轮廓。

被阳光照射的半球的其余部分被三个有着线状纹理的独立区域所占据，并且（在三分之二的情况下）有着不同的反照率。图20从左到右分别是阿尔丁冕状物、因弗内斯冕状物和埃尔西诺冕状物，它们的名字来自莎士比亚戏剧中的地名。国际天文学联合会批准以描述词“冕状物”表示“卵形特征”，但这并不能帮助我们理解它们是什么。与其他地方相比，冕状物的陨击坑较少，而且看起来都很新鲜，所以很明显，冕状物的表面比星球上其他地方的表面要年轻。有相当多的裂缝与因弗内斯冕状物有关，这些裂缝劈开其他地形，形成了一些令人印象深刻的高达10千米的悬崖。

早期的理论认为，每个冕状物都是前一颗因碰撞而破碎的卫星的碎片，现在看来这个理论实在太天真了，我们必须构建一个更复杂的理论。天卫五可能曾经是一个钝化的、布满陨击坑的星球，它的部分表面仍然覆盖在地幔沉积物下面，因而幸存下来。它经历了一次或多次潮汐加热时期，在此期间裂缝的产生和局部的冰火山作用产生了冕状物。冕状物可能被从裂缝中涌出的冰岩浆所覆盖，这些冰岩浆形成的山脊占据了冕状物内的大部分区域。如果在同一时间还发生了爆炸性喷发，这就可以解释在冕状物附近的地形上先前存在的陨击坑为何有柔和的轮廓。

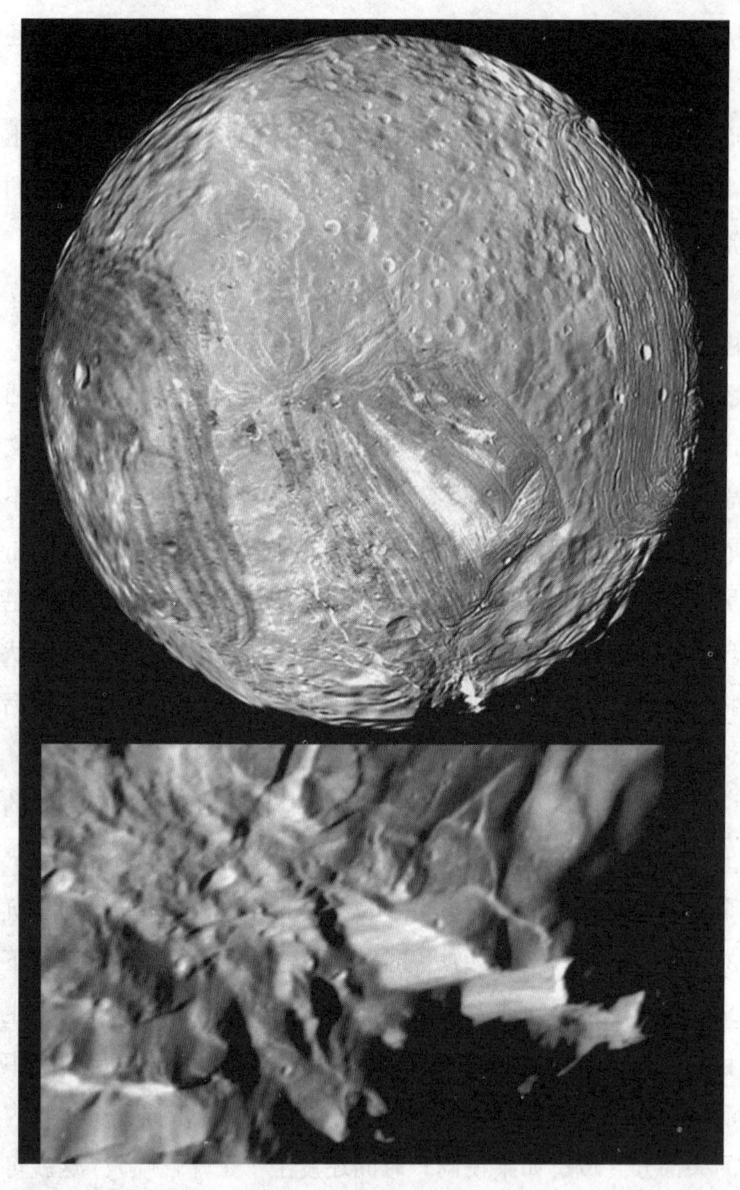

图20 “旅行者2号”看到的天卫五。上侧图片为被阳光照射的半球的拼接图（上图）；下侧图片为稍微旋转了的边缘悬崖的视角，并在36 250千米的范围内以每像素0.7千米的分辨率拍摄得到

这一切至少在表面上是有道理的，尽管当我们最终看到天卫五，特别是还没有被拍到的那半球的更多细节时（目前还没有计划前往天王星的太空任务），它可能会被证明是错误的。潮汐加热可能是由与天卫二的3∶1轨道共振周期，或是由与天卫一的5∶3轨道共振周期驱动的。

氨可以比盐更有效地降低含水液体的冻结温度，同时考虑到在离太阳这么远的地方冰中可能含有大量的氨，天卫一火山喷发出的低温岩浆很可能是一种氨水的混合物，可以在低至零下97摄氏度的温度下从固体冰状态中融化。这比纯水或纯氨的熔点都要低，也是冰的混合物如何形成类似于岩石中硅酸盐矿物混合物的机制的一个例子，其中混合物岩石的熔化温度也低于单质矿物的熔化温度。

107

在天王星的其他规则卫星中，"旅行者2号"在天卫四上发现了断裂的迹象，并在天卫三上发现了明显的断裂特征，但是只有在天卫一上，才发现了断裂作用和冰火山作用的明确证据。

天卫一

天卫一是天王星的第二大卫星，"旅行者2号"在接近天王星的过程中，在12 700千米的距离上看到了它的大致细节，然后将注意力转向了天卫五。它的密度比除土卫六以外的任何土星卫星都要大，而且可能是由一半冰和一半岩石组成（在距离太阳这么远的地方，岩石中可能富含碳）。除了水冰之外，来自地球的光谱研究还探测到了二氧化碳冰，但从许多地区明显经冰火山重塑表面后的性质判断，氨（它没有容易探测到的光谱特征）可能也大量存在（图21）。

108

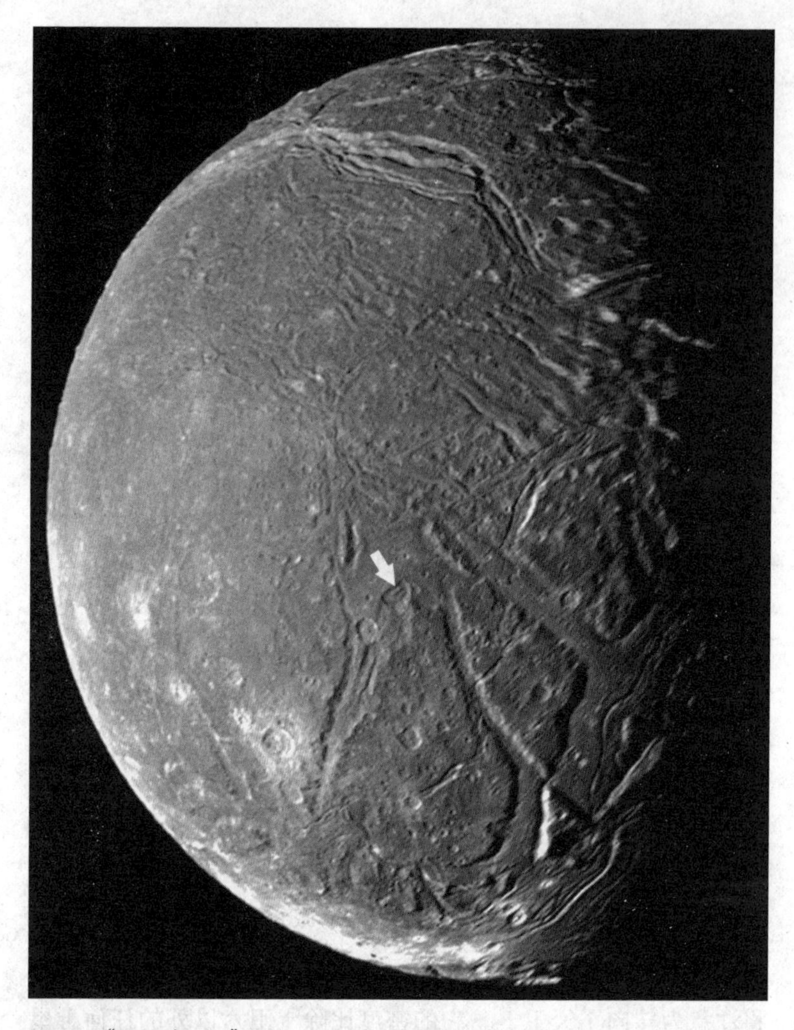

图21 “旅行者2号”从约13 000千米外拍摄的天卫一。箭头标记的是一个陨击坑，它的一部分被从下方和左侧的断层峡谷中流出的黏性冰火山岩浆流所淹没。有一个稍微小一点的相邻陨击坑叠在火山流上，所以它一定是较为年轻的

天卫一的表面有许多的裂缝，是由不同时期一系列断层运动而形成的起伏的条状地形。这种形式的断层是地壳拉伸的表

现，也可能反映了内部加热的一个阶段（可能是潮汐加热，或者，如果足够早的话，可能是放射性加热）导致的内部热膨胀。这些山谷在图21的中间和右下方最宽，可以看到它们底部的陨击坑比两侧高地上的陨击坑更少，这表明它们被什么东西淹没了。从图21中箭头所示的穿过陨击坑的岩浆流的边缘来看，这个“东西”似乎非常黏稠。在月球上，熔岩的黏性要小得多，所以熔岩流会漫延到整个陨击坑，并完全淹没它。在天卫一的低表面重力（不到月球的十分之一）下，融化的氨水混合物会以一种足够黏稠的方式移动，这完全符合条件。

然而，图21右下角附近的宽广的断层谷中间有一条蜿蜒的河道，这个特征似乎与黏性非常高的火山流不太吻合。这让人想起哈德利月溪（图6）和月球上许多类似的特征，它们的来源被认为是快速流动的熔岩造成的热侵蚀或熔岩管道顶部的坍塌——这两种情况都需要低黏度的熔岩。

关于天卫一，我们有太多不了解的地方，它有着复杂的历史，在我的有生之年我们都不太可能解决这些问题。天卫一上喷发过多少种低温岩浆？在地形的断层块之间是否有任何侧向滑动（如果你看图20的右下角，就会发现这一点的暗示）？在左上方黑暗的空间中映出的那些山的剪影是什么？

海卫一

在很多方面，“旅行者2号”都把最好的探测结果留到了最后。它在经过海卫一时距离为2.5万千米，拍摄了其大约40%的表面，并揭示了一个富含多种冰、有一个氮冰构成的极地冰盖的复杂地表（图22）。我在第四章中谈到海卫一的起源可能是一个被捕获的柯伊伯带天体。然而，尽管它的组成成分与冥王星相似，但它在地貌和表面历史方面与冥王星明显不同。

海卫一的氮气大气中含有微量的甲烷，但其含量远低于土卫六大气中的甲烷含量，其表面大气压也只有地球的几十万分之一。虽然这很脆弱，但足以支持在距地表几千米高处一些稀薄的氮冰颗粒形成卷云，同时在30千米高处形成碳氢化合物雾霾层。海卫一的全球平均反照率为0.78，在规则卫星中，只有土卫二的反照率超过了它。这并不是因为它的表面还很年轻（已经发现了近200个陨击坑，直径都不到30千米），而是因为有一层在大气层中冻结的氮霜覆盖在零下230摄氏度的表面。

海卫一有着特殊的季节。虽然它自身的自旋与它围绕海王星的轨道垂直，但这个轨道与行星的赤道的倾角为157度（所以它是逆行的）。最重要的是，海王星的赤道相对于它绕太阳的轨道倾斜了30度。这些影响的结合意味着海卫一的太阳直射纬度在北纬55度和南纬55度之间变化，并且其周期比海王星的165年轨道还要长，因为海卫一的轨道平面也有进动。这意味着海卫一的季节持续时间非常长。当“旅行者2号”飞过时，海卫一的南半球正处于春季的一半，随着氮冰升华并加入到大气中，它的极地冰盖的边缘正在消退。据推测，随着北半球到达秋天并接近冬天，氮从大气中冻结到寒冷的地面上，在黑暗和无法观测的北半球中，极地冰盖正在增长。

透过极地冰盖和较薄的氮霜，我们可以看到它的地表是水冰和二氧化碳冰的混合物，还有甲烷和一氧化碳的痕迹。氨也被怀疑存在，特别是考虑到该地区的冰火山性质，但它尚未被检测到。

图22 “旅行者2号”使用所摄海卫一最好的图像得到的拼接图。成像时，南极冰盖的边缘在南纬5度到20度，该视图的范围从南纬60度左右到北纬40度

海卫一的地形被许多名为“沟脊”（即拉丁文的“凹槽”）的中间有沟壑的山脊所穿过，这让人联想到木卫二上最大的特征，比如图15上的两条相交于图片顶部附近的沟脊。这些沟脊可能

起源于海卫一的轨道变成圆形之前，沟脊在潮汐力的影响下被打开和关闭（或者可能是侧向滑动）。在图22的下半部分，一些沟脊的长度被平坦的平原所掩盖，看起来就像被冰火山熔岩流侵入一般。图22中右侧附近的两个形状不规则的洼地中，也出现了设想中平滑的冰火山岩平原。

图像上部的地形较老。除了一个新沟脊劈开一个旧沟脊的情况，这里的沟脊是不间断的，它们穿过30千米至40千米长坑坑洼洼的地形。没有人知道这是由什么引起的。这被称为“甜瓜形地表”，因为它很像甜瓜的皮，但这并没有多大帮助。这些凹痕可能位于个别“暖冰”从地壳内的洞窟中升起的地方，但不管它们的确切性质如何，它们肯定是另一个冰火山活动的实例。

海卫一的冰火山作用和由沟脊所显示的地壳应力，大概是潮汐作用的结果，在海王星捕获海卫一之后，它的轨道也许用了数十亿年才变成圆形。我们不知道那是多久以前的事，但是散落在星球表面的陨击坑表明这种活动现在已经停止了，或者至少已经减弱了。

然而，海卫一上仍有一种喷发现象。这些现象发生在极地冰盖，“旅行者2号”看到了这一过程（尽管所摄图像质量相当差）。这里所发生的事情似乎是阳光穿过大部分为半透明状态的氮冰，并加热下面较暗的基质。这种热量传导进入氮冰，直到它的底部开始升华。一团氮气泡在冰盖下不断膨胀直到破裂，使气体喷射到天空，带走来自黑色基质的尘埃颗粒。羽状物在数千米的高度上达到浮力平衡。然后它随风飘落，黑色粒子开始沉积，在极地冰盖上留下可以看到的低反照率条纹。如果这个解释是正确的，那么这些是太阳能间歇泉，而不是真正的冰火火山爆发。尽管如此，海卫一仍有资格成为目前已被证实为地质活动仍然活跃的卫星俱乐部的成员，除此之外，该俱乐部还包括木卫一、土卫六和土卫二。

海卫一是一个奇妙的世界，但它不会永远在那里。它的轨道与海王星的距离已经比月球与地球的距离更近了，当它倾斜的轨道经过行星的赤道处时，潮汐过程会将它逐渐拉近。计算表明，海卫一将在大约30亿年内进入海王星的洛希极限，这可能会产生一个远比土星更壮观的环系统。

未来的任务

应该清楚的是，即使是最著名的巨行星卫星，也有更多的东西需要发现，需要更多的任务来传递这些信息。目前，所有太空机构中最早受批准的任务是欧洲航天局的“木星冰卫星探测器计划（JUICE）”，计划于2022年发射，并于2030年抵达木星，进行为期30个月的旅行。首先，利用木卫二和木卫四引力的帮助，探测器将对它们进行一系列近距离飞行，直到最终被捕获到木卫三附近的轨道上，探测器最终将坠入木卫三。此外，还有NASA将在2020年代的某个时候发射的“木卫二任务”（毫无想象力的名字）。和“JUICE”一样，这颗卫星也会绕木星运行，但会对木卫二进行数十次近距离飞行，以研究木卫二的地下海洋，并寻找火山喷发的羽状物。NASA或ESA随后的一项任务（尚未得到资助）可能会在木卫二年轻裂缝旁边的冰层中投放一系列钻孔器或微型“芯片卫星”，以探测冰层并寻找生命。

土卫六也是一个吸引人的目标——NASA正在考虑降落到

土卫六最大的甲烷海洋之一中的提议。我希望这样的任务也能有机会分析土卫二羽状物物质的生物特征。对海卫一和天王星卫星的新探测任务在大多数优先级列表中处于较低的位置，而

114 令人好奇的是，我们只看过它们不到一半的表面。

第六章

火星的卫星：被捕获的小行星

火星是最晚被发现有卫星的行星。在1726年的讽刺小说《格列佛游记》中，乔纳森·斯威夫特这样描述勒皮他飞岛的天文学家们发现火星的两颗小卫星的情形："他们还发现了两颗较小的星星，或者叫卫星，它们围绕火星旋转；最里面的……以10小时的周期在太空中旋转，而后者（最外面的）以21.5小时为周期旋转（第三章第三部分）。"

事实上，在一个半世纪后的1877年，美国的阿萨夫·霍尔使用当时世界上最大的折射望远镜（66厘米口径）进行了长时间的搜索，终于发现了火星的卫星，与斯威夫特的猜测相同，他发现的卫星数量有两个。他首先发现了较小的外部卫星火卫二，六天后又发现了较大的内部卫星火卫一。它们的实际轨道周期（见附录）分别为7.7小时和30.2小时。

斯威夫特关于火星有两颗卫星的猜测很可能是试图使火星符合当时卫星的模式，因为当时认为地球有一颗卫星，木星有四颗已知的卫星，而土星有五颗。1610年，开普勒受到木星的四颗卫星的启发，做了一个类似的数值推测，而当时土星的卫星还没有被发现。斯威夫特给出的短轨道周期的卫星表明了他的想法：如果火星的任何卫星有着更长的周期，它们就会离火星足够远，在现实世界中早就已经被天文学家发现了。

火星的两颗卫星都是潮汐锁定的小岩石天体。火卫一的大小与土星的特洛伊卫星土卫十四（图10）和木卫十五（图12中的巨行星最小的内卫星）差不多。火卫一的尺寸是27千米×22千米×18千米，火卫二的尺寸是15千米×12千米×11千米。它们太小了，自身的重力无法使它们的形状达到流体静力平衡。尽管它们的轨道是顺行的，且只有轻微的偏心率，与火星轨道平面的倾斜度只有1度左右，但它们几乎可以确定是被捕获的小行星。这使得它们更类似于巨行星的不规则卫星，而不是它们的内卫星。它们与月球也没有什么共同之处。

它们的密度还不到水的两倍，密度太低而不可能是坚硬的岩石。虽然光谱研究显示表面没有水合作用的迹象，但内部可能有一些冰。更有可能的是，它们的低密度——这是它们与大多数密度已被确定的小行星的共同属性——是因为在表层风化层之下，它们的内部由松散的碎石块（大小未知）组成。这与土星的土卫六的解释类似，土卫六是一颗冰卫星，但其密度太低，不可能是固体冰。从光谱上看，火卫一和火卫二都类似于且被认为等同于一类被称为碳质球粒陨石的小行星。

火卫一

火卫一的图像（图23）比火卫二被拍摄得到的图像更详细，因为火卫一的轨道高度只有5 645千米（而火卫二的高度为

20 000千米），这使它更接近环绕火星运行的航天器。火星轨道上有几个航天器，它们的的典型高度约为300千米。俄罗斯航天局已经尝试了三次前往火卫一的任务，1988年发射的“火卫一1号”和同年发射的“火卫一2号”中，只有“火卫一2号”成功到达火星，但1989年1月，“火卫一2号”在迫近火卫一并试图着陆时与地球失去了联系。更近一些的是2011年发射的“火卫一-土壤号”，它原本打算从火星表面带回一个样本，但在发射后出现故障，未能离开地球轨道。

火卫一上有许多环形山。年轻的环形山表面很新鲜，古老的环形山表面却很柔和，好像被表土所掩埋。1973年，利用火星探测器“水手9号”发回的火卫一的第一张特写照片，有七个环形山得以命名：其中六个是以包括霍尔在内的研究过火星的天文学家的名字命名的。第七个，也是最大的一个（直径为9千米）被命名为斯蒂克尼，以霍尔的妻子安吉琳·斯蒂克尼的名字命名。在霍尔快要放弃寻找火星的卫星之时，正是她的鼓励使得他最终成功发现了火星的卫星。在图23中，斯蒂克尼环形山几乎占据了火卫一的整个左下角区域，并且离火卫一前半球的中间不远。虽然与火卫一相比很大，但斯蒂克尼几乎不比图4中的月球环形山哥白尼B大多少。它太小了，没有形成一个中心山峰，只有一个简单的碗状环形山。其中还有一个半径为2千米的陨击坑，名叫林姆托克，这是2006年从《格列佛游记》（林姆托克是一个小人国将军的名字）中指定的八个名字之一。

火卫一上的单个环形山肯定是由撞击造成的，但对于可以看到的横跨火卫一表面的100米到200米宽的沟脊，人们存在很大的争议。具体来说，有些是简单的凹槽，而有些则是重叠的

116

117

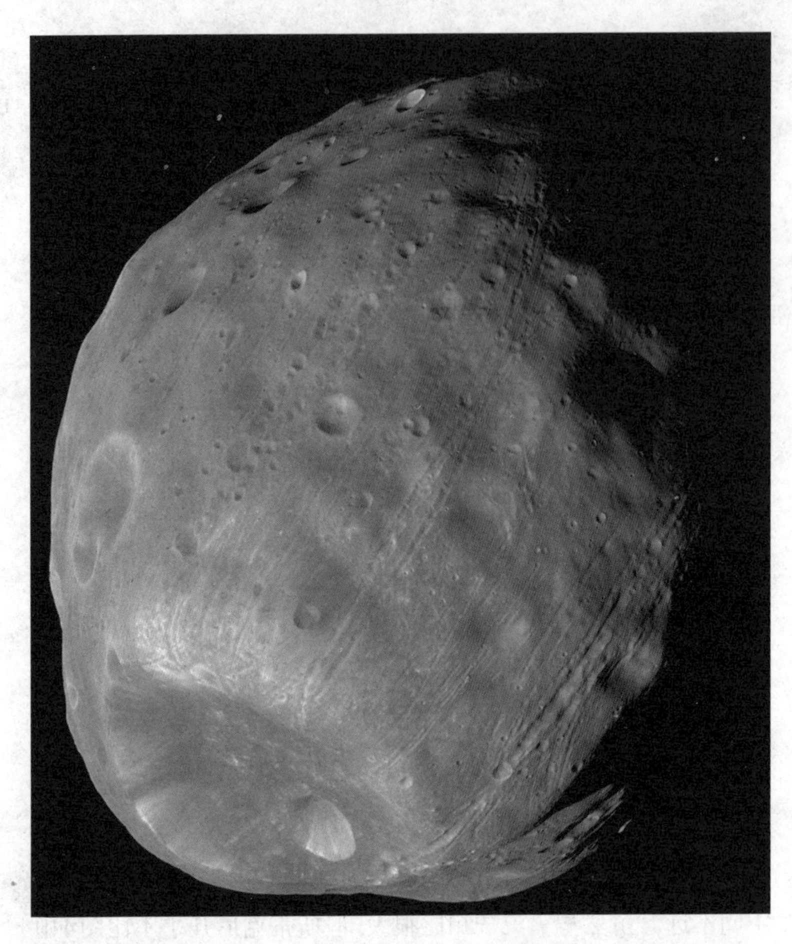

图23　火卫一面向火星的一面，由美国宇航局火星勘测轨道飞行器的"HiRISE"相机在5 800千米范围内以每像素6米的分辨率记录下来

坑链。许多沟脊似乎从斯蒂克尼环形山中向外发散（或者说放射），尽管其中一些穿过了斯蒂克尼环形山的边缘，但它们一定比这个环形山年轻。我们可以识别出不同方向的凹槽类别（特别是图23所示区域以外的凹槽类别），且通过将老旧凹槽与年轻凹槽的重叠，可以显示出一致的年龄关系。

解释这些凹槽的早期假说试图将它们与斯蒂克尼环形山联系起来。它们可能是由斯蒂克尼环形山的喷出物的撞击形成的重叠的次级环形山链吗？这是不可能的，因为任何来自斯蒂克尼环形山的速度若快到足以形成这样的环形山的喷出物，都会从火卫一的低重力中逃逸出来，因此无法回到火卫一表面，而且无论如何，其中一些沟脊都太年轻了。那么它们会不会是斯蒂克尼的撞击造成的裂痕？由于凹槽种类多种多样且有着不同的年龄和方向，这一可能性被否决了。

与斯蒂克尼环形山关联的大多数沟脊的放射状分布，现在被大部分人认为是无关紧要的东西，造成斯蒂克尼环形山的撞击只是火卫一前半球点附近的一个巧合。如果这是正确的，那么真正需要解释的关系是凹槽的模式和火卫一轨道的运行之间的关系。基于此，有两种相互矛盾的假设。其中一种主张，在火星捕获火卫一的过程中，由于潮汐力或来自古代火星大气的空气动力学阻力，在火卫一内部形成的裂缝在表面表现为凹槽。我们观察到的一组断裂的方向表示其可能通过这种方式形成，然后扩大并形成凹槽。这是合理的，但凹槽的直线性所暗示的平整裂缝，很难与火卫一的低密度所要求的碎石和多孔的内部相协调。

另一种基于运动的假说认为，火卫一的低轨道必然会时不时地带着它穿过导致火星上形成巨大环形山的大型撞击所产生的大量喷出物。这些喷出物的碎片可能是在上升或回落时击中火卫一的，这就解释了为什么火卫一从来没有面对过这颗行星的那面上也有凹槽。每一个凹槽都代表火卫一穿过一连串的喷出物，在火卫一表面留下一道损伤痕迹，就像一辆被机枪扫射的汽车的车身一样。在同一次撞击火星时，火卫一会同时遇到几串喷出物，这就是为什么这些凹槽是同一类的。对这种解释也有反对意见，最明显的是，从火星来的喷出物要形成这样的直线凹槽，其弥散程度明显不足（也就是说，一连串喷出物的每个组成部分必须完全沿着相同的方向运动，没有任何横向偏移）。所以，也许最好还是把火卫一的凹槽当作一个尚未得到充分解释的谜。

火卫二

火卫二（图24）上没有凹槽，可能不是因为它没有像斯蒂克尼环形山那样大小的环形山，而是因为它离火星更远。这使得它更不容易受到潮汐力或捕获相关压力的影响，也不太可能遇到火星表面像冰雹一样强烈的喷出物。

火卫二的表面坑坑洼洼，与火卫一上较光滑的部分相似。只有两个环形山得到命名（都是在1973年，基于“水手9号”的图像）：斯威夫特（直径为1千米，由图24中的箭头指向识别）和伏尔泰（直径为1.9千米），后者的特征更加微妙，不太明显，就在图24中斯威夫特环形山的下面。伏尔泰环形山是以法国启蒙运动作家的名字命名的，这位作家比斯威夫特的出生晚四分之一个世纪，他提出了火星一定有两颗卫星的观点。

火星天空中的卫星

火卫一和火卫二是太阳系中除了我们自己仅有的两颗能在天空中看到的卫星：有朝一日，站在火星表面的人类可以在天空中欣赏它们，因为火星是唯一拥有卫星且表面可以登陆的行星。

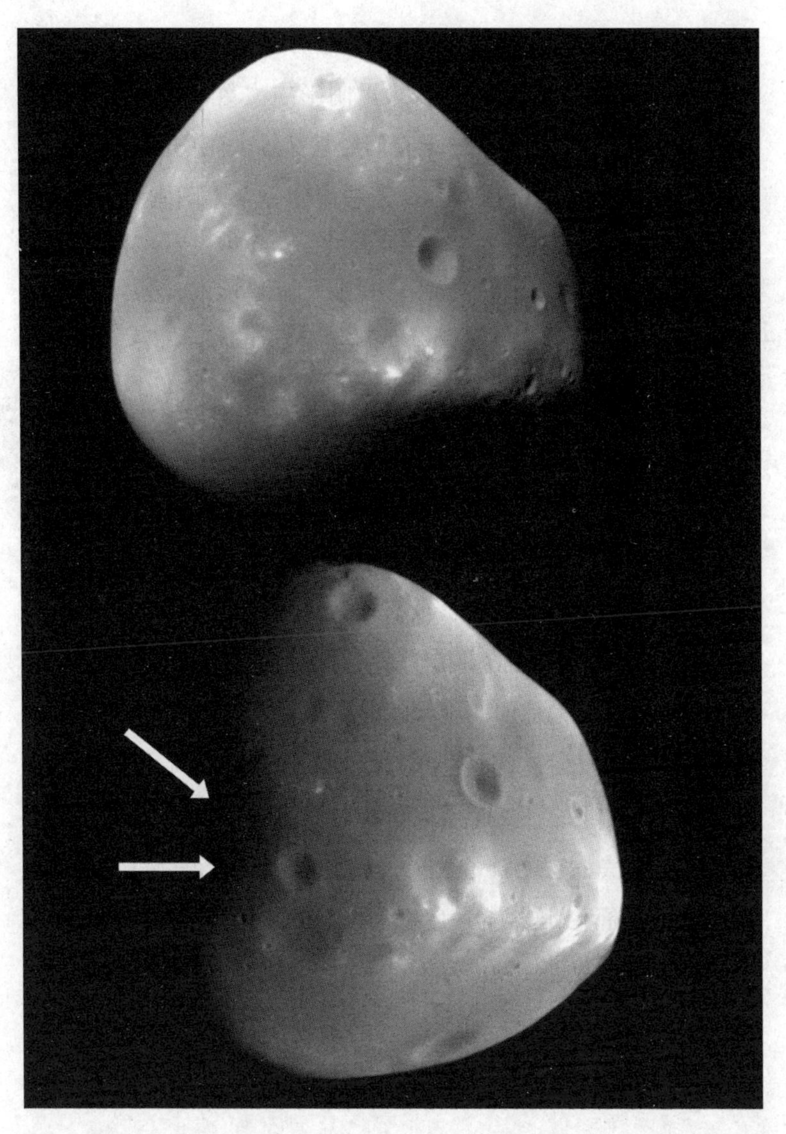

图24　美国宇航局火星勘测轨道飞行器的“HiRISE”相机以大约每像素20米的分辨率记录了两张不同光照下相同的（面向火星的）火卫二表面。斯威夫特环形山由箭头所指处表示

由于它们与火星的距离比月球与地球的距离要近得多，而且位于低倾角轨道上，故它们进入火星阴影的次数比地球上发生月食的次数要多。类似地，白天它们经常穿过太阳的圆盘，但尽管它们离火星很近，但它们太小，无法盖住整个太阳圆盘，所以火星上没有日全食。

当火卫一在头顶上时，它看起来几乎是地球天空中月球的一半大小，但当它接近地平线时，它离观测者的距离几乎变为两倍，因此，这时火卫一看起来相对较小。月球的轨道比地球的半径大得多，因此这种效应对月球来说并不明显。

火卫一穿过火星天空的速度比火卫二快，从火星低纬度地区可以看到火卫一经常从火卫二前面经过，呈现出一颗卫星覆盖另一颗卫星的非凡景象。所有这些现象实际上都被火星表面的火星车携带的相机拍摄了下来，我在拓展阅读中附上了一个火卫一经过火卫二前面的视频的链接。

火卫一和火卫二不会无限期地继续它们的轨道之舞。火卫一离火星如此之近，以至于潮汐力使得它的轨道衰减速度比海卫一还要快。它的轨道上离火星最近的地方正以每年约2厘米的速度下降，以这样的速度，它在被撕裂或撞击到火星表面之前，只能再存活大约5 000万年。

第七章

小天体的卫星

太阳系的小天体可以概括为小行星(集中在但不限于火星和木星轨道之间的岩石或碳质天体);外海王星天体(海王星轨道以外的冰天体,包括冥王星)和彗星(小的冰天体,其轨道高度椭圆化,可以接近太阳)。这三种类型是最有用的分类,但并非所有天体都可以被整齐地分类,因为你很快就会看到第四种类型——半人马型小行星。

在所有这些天体中,只有彗星没有已知的卫星。彗星之所以自古以来就为人所知,只是因为当它们接近太阳时,会形成壮观的气体和尘埃尾巴。最大的彗星的固体部分,或称"彗核",也要比其他任何类型的小天体中最大的天体要小得多。大多数彗星的彗核直径小于10千米,目前还没有直径大到100千米的彗核。

有三个彗核的形状与"相接双星"一致,即两个主彗核在表土下松散地结合在一起。但还没有发现任何彗星拥有一颗围绕其运行的卫星。事实上,如果一颗彗星确实有一颗卫星,它也不太可能存活太久，因为彗星释放出来的气体会使它们彼此的轨道变得混乱。

123

有卫星的小行星

特征明确的小行星会被正式命名，名字前面有一个数字，大致对应于发现的顺序。第一个被发现的小行星是1号小行星谷神星（1801年），它也是最大的小行星，直径为950千米。那些没有被追踪足够长的时间以确定其轨道的小行星没有名字，只有一个由发现年份和两个字母（或两个字母加数字）组成的临时代号。即使是直径约为10米的小行星，当它们接近地球时，雷达也能追踪到它们，但我们可能以后再也看不见它们了。

至2015年，已知184颗小行星有卫星。其中104颗位于火星和木星轨道之间的主小行星带，21颗经过火星轨道内部，55颗与地球轨道相交或接近。它们大多数只有一颗卫星，但其中九颗有两颗卫星。

这一惊人的数字主要归功于过去20年来光学望远镜成像技术的进步。然而，第一颗小行星的卫星是在1994年2月偶然发现的，“伽利略号”在6个月前飞越243号小行星艾达星的照片被下载了下来，它们揭示出一个直径为1.5千米的卫星，后来被命名为艾卫，它布满环形山的表面表明它相当古老。它有一个围绕艾达星的顺行轨道，而且显然与艾达星的组分相似（图25）。

艾卫仍然是唯一一个被近距离观测到的小行星的卫星，要解释它，我们需要考虑艾达星的历史。艾达星是主小行星带的科罗尼斯群的成员，这群小行星拥有相似的轨道，被认为是大约20亿年前两个较大的天体相撞的结果。在科罗尼斯小行星群里

124

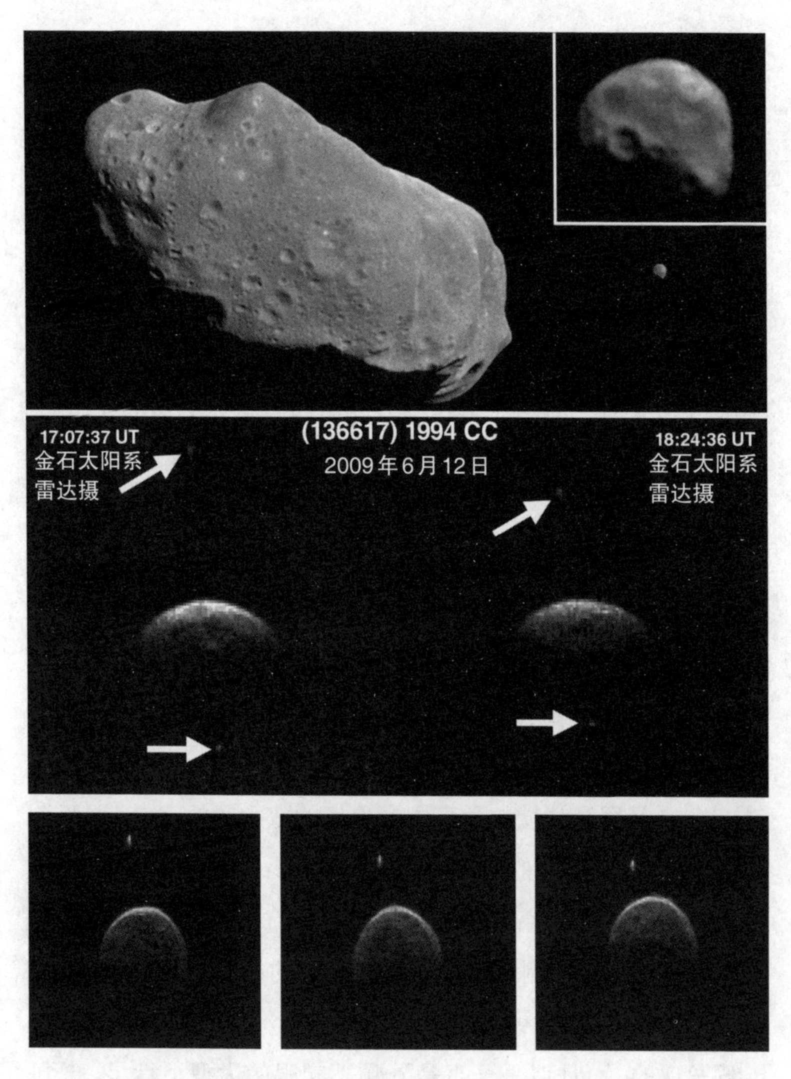

图25 在一张伽利略望远镜拍摄的图片（最上方）中，有小行星艾达星（56千米长）和它1.5千米长的卫星艾卫，框中的图片是在3 900千米范围内放大后的艾卫的最高分辨率视图；中图是2009年雷达拍摄的700米小行星1994 CC和它的两个小卫星（箭头所指），距离约3 000 000千米；最下图拍摄于2015年1月26日，是直径只有1.2千米的2004 BL86小行星及其70米长的卫星的连续三幅雷达图像，拍摄于地球上

有20颗像艾达星这样直径超过20千米的小行星，还有近300颗已知的更小的小行星。艾卫可能是同一个母体在这次碰撞中从艾达星上分离出来的一块碎片，但由于作用力太小而无法独立，或者它也可能是最近才从艾达星身上脱落的一块碎片。

其他所有已知的主带小行星的卫星都是通过望远镜发现的。最早发现于1998年的一个直径为13千米的卫星香卫一（Petit-Prince，意为小王子）在1 184千米的距离上绕直径213千米的45号小行星香女星在接近其赤道面的轨道上以4.8天的周期旋转。2007年宣布发现了一颗直径为6千米、与香女星距离更近的第二颗卫星。这颗卫星还没有正式的名字，但我知道“小公主”这个名字正在等待国际天文学联合会的批准。

在香女星的第二颗卫星被发现之前，人们已经发现了一颗长286千米的87号小行星林神星有着两颗卫星，直径分别为18千米和7千米，在离林神星赤道面很近的地方绕轨道运行，距离分别为707千米和1 357千米。罗慕路斯（林卫一）和雷穆斯（林卫二）是罗马建城的故事中吃狼奶的双胞胎的名字，这两个名字非常合适，因为他们在神话中的母亲是雷亚·西尔维亚，林神星就是以她的名字命名的。

到目前为止的例子都是一颗相当大的小行星，有一颗或两颗小得多的卫星，但90号小行星休神星是不同的。它是在1866年被发现的，而且似乎一直是一颗不起眼的大型小行星，直到2000年才被发现是一颗双星。我们现在知道它是两个天体，大小约86千米，且相距170千米，同步自转并围绕它们共同的质心（即“重心”）公转，周期为16.5小时。由于在大小和质量上无法区分，每一个都可以看作是另一个的卫星。

小得多的小行星只有在接近地球时才能被发现，当它们距离地球不到1 000万千米时，雷达甚至可以对它们进行研究。近地小行星中的16%被证明是双星或三星系统。例如，69230号小行星使神星由两个几乎相同的天体组成，直径约为400米，相距仅1 200米，这两个物体就像是休神星的微缩版。其他的例子类似于缩小版的艾达星/艾卫或林神星/罗慕路斯/雷穆斯。其中，直径为2.8千米的小行星1998 QE_2已被确认其自转周期约5小时，而其600米长的卫星在距离约6千米的轨道上同步自转，公转周期为32小时。小行星1994 CC（图25）有两个更小的卫星。靠内的那颗直径约为110米，在一个半径为1.7千米的轨道上同步自转，公转周期大约为30小时，而靠外那颗（直径约为80米）的自转速度似乎比它的9天公转周期还要快。这是所有已知天体中最小的卫星，直到2015年1月，355米长的小行星2004 BL86以120万千米的距离经过地球，雷达显示它只有70米宽（图25）。

许多小行星的卫星可能是由撞击产生的碎片，但对于较小的小行星来说，它们的卫星可能是随着自转速度的增加，在离心过程中从母体甩出的松散的部分。这种令人惊讶的自转加速之所以会发生，是因为一种被称为“YORP”的效应，即一个小的、不规则的旋转天体的角动量是不平衡的，因为它只有日照面吸收了阳光，而黑暗侧同样会从其温暖的表面发出热辐射。

有卫星的半人马型小行星

半人马型小行星是由冰而不是岩石组成的类小行星，其轨道在木星之外，但在海王星之内。目前已知的半人马型小行星约有200个，其中最大的半人马型小行星比海王星外发现的最大的柯伊伯带天体要小。

127

我们已经通过望远镜拍摄到两颗有时被归类为半人马型小行星的天体的卫星：它们是42355号小行星台神星（162千米）和它的卫星台卫（90千米），以及65489号小行星刻托（200千米）和它的卫星福耳库斯（170千米）。然而，尽管它们在离太阳最近的时候会进入天王星的轨道内部，但它们的轨道是高度椭圆的，在离太阳最远的时候会把它们带到海王星以外的地方，所以它们可能更应该被认为是向内发散的海王星外的天体。

如果要举一个没有争议的可能拥有卫星的半人马型小行星的例子，我们需要转向10199号小行星女凯龙星。这是已知的最大的半人马型小行星，直径约为250千米。2013年6月，人们预测它会正好经过一颗恒星前面（对于一个如此小的天体来说，这是一种罕见的现象），这颗恒星在南美洲的多个天文台都能观测到。当恒星被挡住时，其被称为掩星，精密测量这段时间间隔（通常几秒钟长），为我们提供了最好的方法来计算这个遥远的小天体的大小。因此掩星的路径上的每个望远镜（掩星被遮住的长度与女凯龙星本身的长度相同）都在观测。令所有人惊讶的是，这颗恒星在主要的掩星发生前后分别两次短暂地变暗。唯一可行的解释是，女凯龙星有两个环：一个是7千米宽的环，半径为391千米；另一个是密度较小的3千米宽的环，半径为405千米。

女凯龙星是迄今为止发现的最小的有环的天体。这些环非常狭窄，所以除非它们是在不到几千年前形成的（这将是一个惊人的巧合），否则它们很可能是被几千米大小的、看不见的牧羊人卫星所约束了。

外海王星天体及其卫星

当冥王星最大的卫星在1978年被发现时，冥王星仍然被认为是一颗行星，自从它在1930年被发现以来一直是这样。然而，冥王星一直都是一颗不合格的行星，因为它会暂时在海王星的轨道内部经过，而海王星的质量是冥王星的10000倍。两者围绕太阳的轨道共振为2∶3（冥王星绕太阳公转两圈时，海王星绕太阳公转三圈），但它们永远不会碰撞，因为无论何时当冥王星到达离太阳最近的地方时，海王星要么领先其50度，要么落后其50度。此外，冥王星的轨道是倾斜的，当它最接近太阳时，它会远远“高于”海王星的轨道。事实上，冥王星和海王星之间的距离总是超过地球至太阳距离的17倍。

128

1992年，在海王星外发现了第二个类似冥王星的天体，最终它被命名为136199号小行星阋神星，现在人们认为它的质量比冥王星大27%，尽管体积略小一些。到2015年，外海王星天体总数超过1500个，其中大约80个（包括冥王星和阋神星）已知有卫星。它们中的许多都在类似冥王星的轨道上，距离太阳的距离是地球轨道距离的30倍到50倍。这个区域被称为柯伊伯带，以杰勒德·柯伊伯（1905—1973）的名字命名，他曾预测过类似的现象。再往外是类似物体的“散射盘”，它们的轨道更偏心，也更倾斜，与太阳的距离是地球轨道距离的100倍以上。

冥王星是它们中最大的天体之一，但也并不罕见，它与海王星外的其他天体没有什么区别，只是它于大约60年前第一个被发现。这种关于太阳系的新认识使得继续把冥王星视为行星变得不合逻辑，除非阋神星和其他几个天体也被归为行星。2006年在布拉格举行的国际天文学联合会年会上，人们投票通过了一项对行星的定义，该定义将那些质量不大且共享相似轨道的天体，还有那些轨道与大得多的天体的轨道相交的天体都排除在行星之外。这使得冥王星降级，而其他八颗行星的地位保持不变。

这一决定在某些地方曾经（现在仍然）不受欢迎，但在我看来是正确的，而且相比其他选择减少了许多混乱。那些绕着太阳运行且不符合行星的标准，但其质量足以使它们的引力拉动它们的形状并达到流体静力平衡的天体被归类为矮行星。这一类别包括一颗岩石小行星（1号小行星谷神星），以及至少五颗海王星外的冰天体，冥王星就是其中之一。这一定义听起来合乎逻辑，但在实践中，它并不总是能得到充分的应用，因为只有谷神星和冥王星被清晰地观测到过，以证明它们的形状，而其他的天体只是根据它们估计出的大小和质量，假定它们具有流体静力平衡的形状。

冥王星的第一颗卫星是由美国人詹姆斯·克里斯蒂（1938—　）使用1.55米望远镜在1978年发现的，当时人们只能在冥王星模糊图像的侧面发现一个凸起，尽管更新的望远镜和现代技术很快就能分辨出这两个天体。克里斯蒂提出将这颗卫星命名为“冥卫一”（卡戎），并被国际天文学联合会接受。据说，克里斯蒂想用他妻子夏琳的名字来命名他的发现，夏琳（Charlene）也被称为“夏尔（Char）”。他知道国际天文学联合会不会接受这一命名，但他意识到卡戎（Charon）似乎是合适的，因为在希腊神话中，卡戎是一个摆渡人，负责将死者的灵魂穿过冥河送到冥府，冥府的统治者正是冥王普路托（Pluto，即哈迪斯的罗马名字，也是冥王星的名字）。“卡戎”正统的发音方式是“Kairon”，但克里斯蒂和大多数美国人将其念成“Shairon”或“Sharon”，保留了“Charlene”里的“sh”发音。

目前已知冥王星有五颗卫星（见附录），其中冥卫一是迄今为止最大的一颗。它的质量是冥王星的八分之一，所以这两个天体的质量比地球和月球1∶80的比例要相近得多。正因为如此，它们的重心不在冥王星内部，而是在冥卫一方向，距冥王星表面一个半径的地方。这两个天体相互围绕这一点运行，并被潮汐锁定为同步自转，因此冥王星的一个半球永远看不到冥卫一，而冥卫一的一个半球永远看不到冥王星。当然，这意味着在冥王星上的某个经度会经历永恒的月出或月落。

冥王星的自转轴与轨道的倾斜度为119.6度（所以，像天王星一样，它的自转是逆行的）。它的所有卫星的轨道都在其赤道平面的几分之一度的范围内，而且都是顺行（与冥王星的自转方向相同）且近乎圆形。从最内部（冥卫一）到最外部（冥卫三）的轨道周期接近（但不完全为）1∶3∶4∶5∶6共振。

较小的卫星的名字以普路托（冥王星）和卡戎引导的冥界主题命名：倪克斯（冥卫二），黑暗和夜晚的女神，是卡戎的母亲（摆渡人而非克里斯蒂夫人的那位卡戒）；海德拉（冥卫三）是一条九头蛇，有人说这是冥王星暂时占据第九大行星之位的一个暗示；斯堤克斯（即冥河，冥卫五）是以冥河女神的名字命名的，冥河是摆渡人卡戎经常横渡的那条河；刻耳柏洛斯（冥卫四）是守护冥界入口的三头犬。2015年7月，“新视野号”探测器飞行中发现的所有其他卫星都将会有类似的冥界名称，但它没有发现任何卫星。然而，在一次短暂的相遇中，它以14千米/秒的速度从距离冥王星约1万千米、距离冥卫一约2.7万千米的地方经过，清晰地展示了它们全球的大片区域，最高分辨率时的像素尺寸不到200米。

冥卫一的形成可能是一次类似于月球起源的“巨大撞击”事件的结果，在这种情况下，这四颗较小的卫星可能是同一事件产生的正向外迁移的碎片。如果冥卫二和冥卫三在同步自转，这将表明已经发生了卫星外移，因为计算表明，潮汐力不足以在它们当前的距离实现这一目标。如果它们不是在同步自转，这将暗示捕获（伴随着捕获的所有困难）是一个更有可能性的起源。

冥王星被氮、甲烷和一氧化碳的冰覆盖着，覆盖在由水冰构成的更坚固、更不易挥发的“基岩”上，但其密度足以让它的内部深处存在岩石。它的大气来源于表面的冰，主要成分为氮气。在所有这些方面，它看起来都很像海王星的大卫星海卫一，你可能还记得，海卫一很可能是从柯伊伯带捕获而来的。“新视野号”的特写图像发现冥王星的表面环形山比海卫一的还要少，这表明它的表面直到最近还在不断更新。一些高分辨率的图像似乎根本不包含环形山，这表明该地区的地表年龄不到1亿年，不存在类似海卫一上的沟脊和甜瓜形地表。相反，在一些地区，水冰从其他冰中伸出，形成锯齿状的3千米高的山脉区域，还有一氧化碳冰形成的平原，其表面看起来像地球北极和火星部分地区的冻融过程形成的更大尺寸的地表。

相比之下，冥卫一的表面主要由水冰组成，但在“新视野号”

之前，望远镜就探测到了水合氨，这引发了人们对最近活跃的冰火山活动的猜测。人们对冥王星-冥卫一系统的潮汐加热程度还不清楚，但太阳对大倾角的冥王星-冥卫一轨道的拉力变化存在影响，而冥卫一比冥王星更脆弱，可能足以产生部分融化内部混合物冰所需的热量。“新视野号”团队的第一个想法是，形成现在的冥王星地形所需的热量无法从潮汐角度来解释。虽然他们不能举出任何可行的替代机制，但他们的推测也使人们对用潮汐加热来解释在其他冰天体上的年轻表面产生了怀疑。一种更简单但同样有争议的解释是，冥卫一是在相对较近的过去被捕获的，或者是由一次最近的，而不是在数十亿年前的巨大撞击形成的。

冥卫一也有一个年轻的表面（图26），尽管它最近的活动迹象比冥王星少。值得注意的是，它有几千米深的直边沟脊，深达几千米，类似于天王星的卫星天卫一（图21）和天卫三。冥卫一上还有一个被“新视野号”团队命名为“魔多区域”的黑色北极冰盖。

除了绘制冥王星和冥卫一的地形和组成，“新视野号”还在太阳和地球被这两颗天体依次遮挡的情况下，对它们进行了回望，以了解冥王星的大气层，并试图确定冥卫一是否存在大气层。这些较小的卫星被拍摄得足够好，可以显示出它们的形状，和预期的一样，它们是不规则的。

2018年或2019年，“新视野号”将飞越至少另一个柯伊伯带天体及其卫星（如果有的话）。一般来说，海王星外天体的卫星有很多组合，它们的起源很可能是碰撞。136199号小行星阋神星（Eris）的名字是古希腊代表冲突与不和女神的厄里斯，它

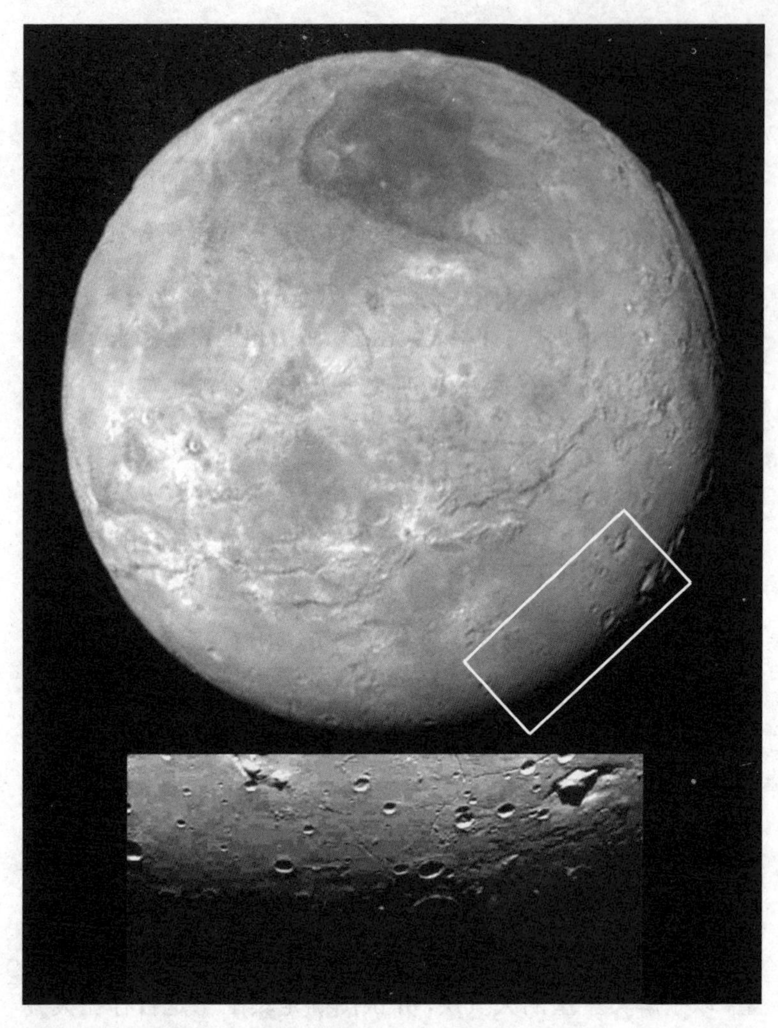

图26 “新视野号”拍摄的冥卫一的全球视图。注意靠近东部边缘(右手边)的南北裂缝系统和靠近赤道的第二个主要裂缝系统。黑暗的极地地区是魔多地区。明暗分界线附近的撞击坑由其内部阴影显现出来。在其他地方,它们的显现是因为周围有明亮的喷出物沉积,或者有时是因为它们挖掘出了较暗的物质。该框指示的区域显示在下面的更详细的图像中(记录在几个小时后,因此明暗线已经迁移)。这张图显示了一个被壕沟环绕的令人费解的山峰,在分辨率的限制下,有几个直裂缝,让人想起图5中扬松E西部的裂缝。在那里,它们是由垂直岩浆幕(岩脉)的浅层侵入造成的月海玄武岩的伸展性裂缝,而在冥卫一上,它们可能是由水–氨混合物组成的低温岩浆的类似侵入造成的

的发现引发了关于冥王星地位的混乱。在2005年发现了一颗它的已知的卫星，并被命名为戴丝诺米娅（阋卫一），由神话中厄里斯的女儿所生。阋神星的半径约为1 200千米，而根据反照率，阋卫一的半径一定在100千米至300千米的范围内。阋卫一的轨道证明，阋神星的质量明显比冥王星大（约27%），尽管它的体积略小一些。阋神星和阋卫一是观测起来具有挑战性的天体，因为它们目前与太阳的距离是冥王星与太阳的距离的3倍。

对这些遥远天体的大小估计是非常不准的，因为它们大多太小、太远，无法在望远镜中显示出可分辨、可测量的圆盘。如果观测到一颗恒星的掩星并测量了其持续时间，你可以使用该天体的轨道速度来推断该天体在恒星前面经过的那一部分的大小，尽管那很可能只是一个弦而不是一个完整的直径。这些都是罕见的事件，这就是为什么天文台做足了准备观测10199号小行星女凯龙星，并碰巧发现了它的光环。在没有任何掩星数据的情况下，如果你假设一个反照率，你可以用一个物体的亮度来推断其大小。这种方法的不确定性意味着，我们有一段时间不能确定阋神星比冥王星大还是小，尽管我们已经确定阋神星的质量更大。

冥王星和冥卫一各自的大小最初是在1985年到1990年之间发生的一系列相互掩星事件中推断出来的，当时它们的轨道平面穿过了地球的视线。2014年7月至2018年10月之间发生的一系列类似事件，应该能大概揭示出柯伊伯带天体385446号小行星蓝神星的真实大小，该天体长约80千米，其卫星梭隆多长约50千米（它们的名字来源于J. R. R. 托尔金的作品《魔戒》系列）。这不仅仅是一时的兴趣，因为目前对蓝神星的大小的最佳估计意味着它的密度小于水的密度，在这种情况下，它一定是一个冰碎石团。蓝神星绕太阳的轨道与海王星的共振率为4∶7，而冥王星与海王星的共振率为2∶3。

柯伊伯带最大的天体之一是136108号小行星妊神星，其平均半径约为650千米。它的亮度在不到两个小时的时间里上下变化，显示出它是一个细长的天体，自转周期为3.9小时。这种极快的自转被认为使它扭曲成一个细长的椭球体，而不是像自转较慢的物体那样在赤道处凸起。它的两极直径可能略低于1 000千米，而横跨赤道的最大和最小直径分别为1 960千米和1 520千米。它的密度足以拥有岩石的核心，但它的表面光谱显示出是结晶的水冰。

妊神星有两颗已知的卫星，妊卫一（直径为340千米，轨道位置为49 900千米，周期49.5天）和妊卫二（直径为170千米，轨道位置为25 700千米，周期18.3天）。就像妊神星本身一样，它们也具有结晶水冰的光谱特性，而且可能具有高反照率。由于它们具有共同的特性，因此不太可能是被捕获而来的。妊神星的卫星很可能是在一段更快速的自转期间或一次碰撞中被抛出的碎片。妊神星（哈乌美亚）是以夏威夷的一位女神的名字命名的，她的许多孩子（包括喜雅卡，即妊卫一；那玛卡，即妊卫二）都是从她身体的不同部位孕育出来的，就像有着相同名字的妊神星一样。

除了妊神星外，唯一的海王星外的三星系统有着一个类似冥王星的轨道（与海王星的共振为2∶3），被称为（47171）1999 TC_{36}，它尚未被正式命名。它由一个中心双星组成，每一个组成部分的直径都是260千米。所以和休神星一样，我们面临着一个两难的境地，那就是确定谁是谁的卫星。然而，在这种情况下，这对双星还有一颗140千米长的卫星围绕着运行。这两个双星的平均密度，在没有通过相互掩星来确定其大小的情况下来估计，似乎只有冰的三分之二左右。如果这是正确的，那么它们两个可能都是一堆冰冷的碎石。

136

第八章

其他行星系统中的卫星：系外卫星

第一次明确发现围绕另一颗恒星的行星（“系外行星”）是在1995年。20年过去了，我们知道有1 000多颗恒星有系外行星，其中近一半有一颗以上的系外行星。现在看来，有20%的类太阳恒星至少有一颗系外巨行星，而至少有40%的类太阳恒星可能有质量较小的系外行星。

在我们的太阳系中，卫星比行星要多得多，如果系外卫星的数量少于系外行星的数量，那将是令人惊讶的。然而，探测它们将是非常具有挑战性的。只有少数特殊的系外行星通过直接成像被发现，而且目前任何系外卫星都远低于可见阈值。绝大多数系外行星都是通过恒星围绕自身和系外行星之间的重心旋转时径向速度的周期性变化来推断的，或者，如果系外行星的轨道平面在我们的视线范围内，还可以通过观察系外行星穿过恒星圆盘时星光的微小下降，即遮挡了一小部分恒星的光线来推断。

在这两种方法中，仔细地使用凌日法是发现系外卫星的最大希望。如果系外卫星在系外行星之前或之后从恒星前面经过，这将导致恒星亮度在行星本身造成的更大的亮度下降之前或之后，出现轻微下降。此外，即使出现因系外卫星凌日引起恒星亮度的下降而无法被检测到的状况，对一系列重复的系外行星凌日的分析可能会揭示它们的精确时间波动，这是系外行星移动到系外行星-系外卫星的重心的另一侧引起的。一个类似于木卫一的系外卫星可以通过等离子体环的存在来发出信号，等离子体环面穿过这颗恒星的前面，既可以使其光线变暗，也可以通过特征光谱线显现出自身。也可能存在可识别的无线电波发射，能够追踪到这样的等离子体环面。也许关于系外卫星最有力的证据是一颗名为J1407的橙矮星的行星，它被发现是一颗“超级土星”，有一个巨大而壮观的光环系统，当光环挡住恒星时，恒星的光线会出现多次下降。环上的空隙强烈地暗示着其与大卫星的共振，或者存在质量较小的牧羊犬卫星。

到目前为止，这些技术都没有提供确凿的证据证明系外卫星存在。在系外行星研究领域工作的同事们向我保证，我们可能离成功不远了，尽管我认为第一次真正探测到一颗系外卫星后可能还需要一段时间才能排除合理的怀疑并加以确认。

为什么系外行星很重要？这么说吧，在我们的太阳系中，至少有很多看起来适合居住的卫星（木卫二和土卫二），就像宜居行星（地球和火星）一样多。如果海底的热液喷口真的像第五章中提到的那样，是生命起源的好地方，那么全银河系中有着内部海洋的冰系外卫星都可能存在生命。正如我们所看到的，即使体表温度会远低于水的冰点，这种生命存在的可能性也是合理的。

潮汐加热的冰系外卫星可以在远离恒星的“宜居带”之外孕育生命，这需要卫星表面有液态水。这些生命可能主要为微生物生命，也许有人会说，任何智能的多细胞生命都很难在水下发展出技术文明，但可能存在生物的卫星比可能存在生物的行星要多得多这一点是可以确定的。

内部海洋并不是我们所能想象的系外卫星上的唯一宜居环境。对系外行星的研究发现，许多系外巨行星离恒星的距离比木星离太阳的距离要近得多。这样一颗系外的巨行星可能有一颗类地系外卫星，在其上可能会出现类地生命，甚至智慧生命。近期科幻作品中一个著名的例子是潘多拉星球，詹姆斯·卡梅隆的电影《阿凡达》就是以潘多拉星球为背景的，它在其恒星的传统宜居带绕着一颗系外巨行星运行。

也许，如果我们真的与外星人建立了联系，他们可能会怀疑我们是来自一个围绕着巨行星运行的世界，而不是来自一个围绕着恒星运行的世界。

附录：卫星数据

这个表格包含了所有已知的行星和冥王星的卫星数据。为了利用单个相同的数据来比较非球形卫星的大小，此处使用卫星的平均半径。

地球与火星的卫星

名称	轨道半径（10^3千米）	轨道周期（天）	平均半径（千米）	质量（10^{20}千克）	密度（10^3千克/立方米）
月球	384.4	29.53	1 737	734.2	3.344
火卫一	9.378	0.319	11.2	0.000 106	1.90
火卫二	23.46	1.26	6.1	0.000 024	1.75

木星的卫星

名称	轨道半径（10^3千米）	轨道周期（天）	平均半径（千米）	质量（10^{20}千克）	密度（10^3千克/立方米）
四颗内卫星	128—222	0.29—0.67	10—73	<0.075	<3.10
木卫一	421.6	1.769	1 822	893	3.50
木卫二	670.9	3.551	1 561	480	3.01
木卫三	1 070	14.97	1 737	1 482	1.94
木卫四	1 883	26.33	2 410	1 076	1.83
五十九颗外围卫星	7 507—24 540	130—779	1—85	<0.095	

土星的卫星

名称	轨道半径（10^3千米）	轨道周期（天）	平均半径（千米）	质量（10^{20}千克）	密度（10^3千克/立方米）
十六颗内卫星与共轨道卫星	134—377	0.56—2.74	16—93	<0.019	<1.300
土卫一	185.5	0.952	197	0.379	1.15
土卫二	238.0	1.37	251	1.08	1.61
土卫三	294.7	1.89	528	6.18	0.985
土卫四	277.4	2.74	561	11.0	1.48
土卫五	572.0	4.52	763	23.1	1.24
土卫六	1 222	15.9	2 575	1 346	1.88
土卫七	1 481	21.3	133	0.056	0.55
土卫八	3 561	79.3	746	18.1	1.09
三十八颗外围卫星	11 110—25 110	449—1 491	3—109	<0.083	

天王星的卫星

名称	轨道半径（10^3千米）	轨道周期（天）	平均半径（千米）	质量（10^{20}千克）	密度（10^3千克/立方米）
十三颗内卫星	49.8—97.7	0.345—0.923	5—81		
天卫五	129.4	1.41	234	0.66	1.20
天卫一	191.0	2.52	578	13.5	1.67
天卫二	266.3	4.14	528	585	1.40
天卫三	435.9	8.71	561	789	1.71
天卫四	583.5	13.5	763	761	1.24
九颗外围卫星	4 276—20 900	267—2 824	9—75		

海王星的卫星

名称	轨道半径（10^3千米）	轨道周期（天）	平均半径（千米）	质量（10^{20}千克）	密度（10^3千克/立方米）
六颗内卫星	48.2—105	0.294—0.950	18—102		
海卫八	117.6	1.12	208	0.5	
海卫一	354.8	5.88	1 353	214	2.059
海卫二	5513	360	170		
五颗外围卫星	15 730—48 390	1 880—9 374	20—30		

冥王星的卫星（一些小卫星的平均半径为估算的范围）

名称	轨道半径（10^3千米）	轨道周期（天）	平均半径（千米）	质量（10^{20}千克）	密度（10^3千克/立方米）
冥卫一	19.6	6.39	608	16.2	1.85
冥卫五	42	20.2	5—13		
冥卫二	48.7	24.9	21×18		
冥卫四	59.0	32.1	6—17		
冥卫三	68.8	38.2	27×20		

索　引

（条目后的数字为原书页码，见本书边码。注：特定于月球的条目列在“月球”下。与其他天体的卫星有关的一般项目可以在“卫星”条目下找到。参见每个卫星的单独名称。）

D

E

G

H

I

J

K

L

M

N

O

P

U

V

W

Y

David A. Rothery

MOONS

A Very Short Introduction

Contents

List of illustrations

Chapter 1
The discovery and significance of moons

It would be meaningless to write about the discovery of our moon, the one that goes round the Earth—'the Moon', as I shall call it from now on because that's its name. The Moon is almost as obvious as the Sun. Clouds permitting, we can see it in the evening sky nearly half the time. If we are awake early, there it is in the pre-dawn sky most of the rest of the time. It can often be spotted in daylight too.

Humans have been well aware of the Moon for as far back as records go, but surely before then too because the Moon must often have been a welcome source of illumination at night. Possibly the oldest Moon-related artefacts are 30,000-year-old bone plates engraved with dots or lines, thought by some to be a way of keeping track of the Moon's phases, as it swells from new to full and then shrinks back from full to new on its 29.5-day cycle. The Moon's appearance changes in this way because its orbit carries it round the Earth, continually changing its position in the sky relative to the Sun so that the amount of the illuminated hemisphere we can see changes too.

It would be equally meaningless to try to pin down who first realized that the Moon goes round the Earth. To most ancient peoples it must have seemed pretty obvious that *all* the heavenly bodies go round the Earth. In fact that's wrong, and the Moon is actually the *only* natural body in the sky to go round the Earth. I'm not referring to the daily (24-hour) motion whereby the Sun and all other celestial bodies rise and set—that's just an apparent motion because the Earth is spinning—but to the Moon's 29.5-day journey round the sky relative to the Sun. This 29.5-day period is a combination of the 27.3 days that it takes the Moon to make a 360° circuit round the Earth plus just over two days more to compensate for the fact that the Earth has moved about a twelfth of the way round the Sun in the meantime.

In due course, the discovery of moons orbiting another planet showed that the Earth is not special so far as motion is concerned. This provided crucial evidence to help overturn deeply entrenched 16th- and 17th-century orthodoxy that held the Earth to be the hub of all creation.

Although some ancient Greek philosophers had preferred the notion that the Earth and planets move round a central fire (the Sun), they were in a minority. By the start of the 17th century, the long-established view of the universe had the Earth at the middle, with the known heavenly bodies going around it. The Moon was rightly believed to be closest, then Mercury, Venus, the Sun, Mars, Jupiter, and Saturn. Beyond this was a sphere bearing the stars. So far as most philosophers were concerned this sphere too rotated, because the Earth seemed motionless, although some Greeks and Indians had suggested that the Earth was spinning.

Greek philosophers and their later followers were wedded to the concept that heavens should be 'perfect', so they tried to interpret the observed movements of heavenly bodies in terms of uniform circular motion. As the precision of observations improved, more and more flaws appeared when trying to fit theory to observation. This led to the development of elaborate and cumbersome explanations, in which smaller circles were fitted on to larger circles in a system of 'epicycles', and also in which the speed of motion was uniform only when measured about a special point that did not coincide with the centre of the relevant circular track.

This elaborate Earth-centred (geocentric) view of the cosmos is generally referred to as the Ptolemaic system, after a Greco-Egyptian named Claudius Ptolemy who worked in Alexandria about AD 150, and was endorsed by the Catholic Church. In much of Europe it was dangerous to promote a contrary view, and geocentric ideas also held sway in China and across the Islamic world.

However, in the early 16th century the Polish astronomer Nicolaus Copernicus (1473–1543) developed a rival theory, which had the sphere of fixed stars surrounding the planets including the Earth that went round the Sun (a heliocentric model), and only the Moon going round the Earth. This is essentially correct (except that the stars are not fixed to a sphere—they are merely very distant), but to make it fit the available observations Copernicus had to invoke even more epicycles than were needed in the Ptolemaic system. It wasn't until Johannes Kepler (1571–1630) introduced elliptical, rather than circular, orbits round the Sun in 1609 that the heliocentric model became more elegant. It took until 1687 for Isaac Newton (1643–1727) with his laws of motion and theory of gravity to explain *why* orbits are ellipses.

Although Copernicus's model became known to many colleagues, he was reluctant to publish. It was only in the year of his death, 1543, that his great work *De revolutionibus orbium coelestium* (*On the Revolutions of the Heavenly Spheres*) was published. Contrary to popular myth, this was not immediately banned by the Church, but its views were certainly controversial because they contradicted the biblical view of the cosmos with the Earth as the centre of creation.

Enter Galileo Galilei (1564–1642), an Italian scientist working in Padua. In 1610 he turned one of the world's first telescopes (which he had built himself) skywards. As well as discovering the phases

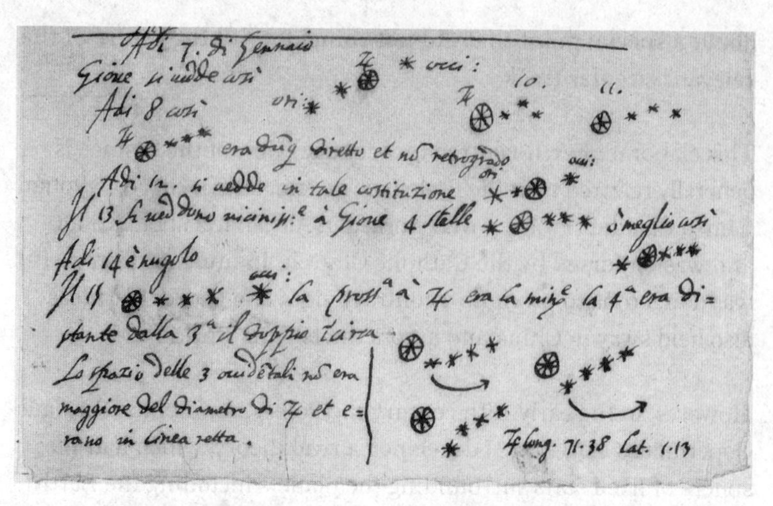

1. Part of a page of notes by Galileo, documenting four moons of Jupiter on successive nights from 7 to 15 January, 1610 (except 14 January, which was cloudy).

of Venus and a multitude of faint and otherwise invisible stars in the Milky Way, he saw up to four faint 'stars' accompanying Jupiter (Figure 1). He was able to note that these shifted back and forth to either side of Jupiter, and concluded after only a few nights' observations that these objects were orbiting Jupiter in a plane very close to his line of sight.

Galileo disseminated his observations in March 1610 in a pamphlet that he called *Sidereus Nuncius* (*Starry Messenger* or *Starry Message*). That autumn, when Jupiter reappeared in the sky, several astronomers, including the Englishman Thomas Harriot (1560–1621) and the German Simon Marius (1573–1625), confirmed the existence of these four moons of Jupiter. In fact Marius claimed that he had discovered them himself in 1609.

They were described as 'satellites' (still a correct alternative to 'moons') before they were described as 'moons'. Galileo himself used neither term (at least at first), referring to them as 'stars', though we can take this as a reference to their point-like appearance rather than an indication that he thought they were the same as conventional stars. Marius called them 'Jovian planets' in his 1614 work *Mundus Jovialis* and Harriot also wrote of his own experience of seeing 'the new planets'. It was Kepler who first called them satellites, when writing of them in 1610, adopting a Latin word meaning 'one who attends a more important person'.

Irrespective of whether there was any truth in Marius's claim, Galileo was the first to publish, and so is credited with the first observational proof that all celestial motion is not centred on the Earth, nor indeed all on the Sun.

Making use of moons

Jupiter's moons orbit their planet in a predictable, clockwork fashion. However, within fifty years of their discovery the Italian Giovanni Cassini (1625–1712), working in Paris, had noted some slight but systematic discrepancies in the regularity with which the moons disappeared into Jupiter's shadow. The intervals between disappearances (eclipses) become slightly shorter when the Earth's orbit is bringing it closer to Jupiter, and lengthen slightly as the two planets draw apart. Cassini rightly suggested that this is because light does not travel infinitely fast, as was then believed, but in fact takes a measurable time to get from Jupiter to the Earth. In 1676 he argued that it takes about ten or eleven minutes for light to cross the Earth–Sun distance, which was pretty close to the correct value of eight minutes and thirty-two seconds. His assistant Ole Rømer (1644–1710) soon came up with observations that enabled a more precise estimate, but it was about fifty years before this moon-based demonstration of the finite speed of light became generally accepted. It also needed a good measurement of the Earth–Sun distance before this speed could be confidently converted into familiar units, though in fact Cassini had made an estimate of the Earth–Sun distance in 1672 that was only about 7 per cent below the correct value of 150 million km.

Telescopes had to improve beyond the rather crude devices built by Galileo and his contemporaries before other moons could be discovered. The Dutchman Christiaan Huygens (1629–95) built a better telescope, and in March 1655 he found the first known moon of Saturn, referring to it as *Luna Saturni* (Saturn's moon) and in so doing probably became the first to use the term 'moon' to refer to the satellite of another planet. Cassini found another four fainter (and therefore smaller) moons of Saturn between the years 1671 and 1686.

To some at the time there seemed to be a logic in Saturn having five moons whereas Jupiter had only four. However, many more discoveries were to come and these destroyed any semblance of a simple pattern. In 1781, the planet Uranus was discovered by William Herschel (1738–1822), a Hanoverian working in England, and in 1787 he discovered two of its moons followed by two more moons of Saturn in 1789. The next moon to be seen was the largest moon of Neptune, found by the Englishman William Lassell (1799–1880) just seventeen days after the planet itself had been discovered in 1846. Lassell is jointly credited with the American father and son William (1789–1859) and George (1825–65) Bond for the discovery of an eighth moon of Saturn in 1848, and individually discovered two more moons of Uranus in 1851.

The two tiny moons of Mars (the only ones it has) were discovered in 1877 by the American Asaph Hall (1829–1907). A fifth moon of Jupiter was discovered by the American Edward Emerson Barnard (1857–1923) in 1892, nearly 200 years after Galileo had found the first four. The delay was because this one is much smaller and also closer to Jupiter, which are both factors that make it hard to see. It was the last moon to be discovered visually. Subsequent discoveries were by photography, the first being a ninth moon of Saturn found by the American William Pickering (1858–1938) in 1899 on plates that had been taken the previous year in Peru. More recently discoveries have been achieved by digital cameras fitted to telescopes or on spacecraft.

So, there are lots of moons in our Solar System—in fact far more than I have yet mentioned—but is knowing about them of any use? Perhaps surprisingly, the moons discovered by Galileo found a practical use even in the 17th century. Until chronometers (clocks) had been developed that could keep accurate time despite being transported (achieved in the late 18th century), it was immensely difficult to determine longitude. On land you could try to measure the distance between places, whereas at sea the best you could do was to estimate distance travelled from course and speed. This sometimes went disastrously wrong, as in 1707 when a homeward-bound British naval fleet struck rocks off the Isles of Scilly with the loss of more than 1,400 lives.

The independent clockwork of Jupiter's moons offered a partial solution, especially after Cassini produced a set of tables accurately predicting the times of their eclipses. All you had to do to fix your longitude was to measure the time between a particular eclipse and local noon at your location, and then use Cassini's tables to decide how far east or west you were from Cassini's reference longitude.

Cassini himself secured a commission from Louis XIV of France to improve the maps of his kingdom, and in the 1670s he and his team set about determining the longitudes of the major cities of France, relative to Paris. They would observe an eclipse of one of Jupiter's moons and then use a pendulum clock (a perfectly fine chronometer if we don't try to move it) to measure how long this occurred before local noon (determined by observing when the Sun reached its peak altitude).

Sometimes the truth is hard to bear. Many cities were found to be up to 100 km closer to Paris than expected, showing France's east–west extent to be less than previously believed. Louis reputedly complained that his own astronomers had deprived him of more territory than his enemies ever had.

Cassini successfully exported his technique overseas. The difficulties in keeping a telescope trained on Jupiter from a moving ship made it impractical while actually at sea, but it was fine for determining longitude ashore, and indeed was still being used in the early 19th century as an aid to mapping in the American West.

The Moon itself provided another means of determining longitude, because it moves across the sky by an amount roughly equal to its own diameter every hour. By measuring the position of the Moon relative to a known star we can determine the time, but we have to correct for parallax because the Moon is sufficiently close to the Earth for its position in the sky relative to the background stars to be slightly different when seen from different locations. The principles of the method were proposed by the Englishman Edmond Halley (1656–1742), later of comet fame, around 1683. The necessary measurements required precise determination of the angle between the Moon and a star followed by difficult calculations that could take over half an hour to perform, until some tables were developed that reduced this to about ten minutes. This method of using the angular separation between the Moon and reference stars was known as 'lunars' and was widely used at sea between 1767 and about 1850, by which time reliable marine chronometers had become affordable. However, it continued to be taught to navigators as a back-up technique for at least a further fifty years.

Another use of moons is to measure the mass of the object about which they orbit (or rather the combined mass of the two objects). This is because the square of the orbital period is proportional to the cube of the orbital radius divided by the sum of the masses. If the moon is much smaller than the planet, then its own mass is negligible, so the orbital period tells us the planet's mass. To turn this measurement into familiar units, we need to divide the number we get for the mass by a number that is related to the 'gravitational constant' *G* (which basically tells us how much mass it takes to produce a given strength of gravity).

Had *G* been known early enough, then scientists could have used the Moon's orbital period to weigh the Earth. It didn't work out that way, because *G* was first adequately determined in 1798 by the Englishman Henry Cavendish (1731–1810) in an experiment that essentially determined *G* and the mass of the Earth at the same time. However, once *G* was known and the scale of the Solar System had been determined (Cassini had not been far out, and the uncertainty had been considerably reduced a hundred years later thanks to observations of transits of Venus across the Sun in 1761 and 1769), the orbital periods of other planets' moons could be used to determine the planets' masses and their densities. This revealed, for example, the enormous masses of Jupiter and Saturn (318 and 95 times the Earth's mass, respectively) and also their low densities (Saturn has only 69 per cent the density of water).

Venus and Mercury have no moons, and so their masses had to be estimated from the tiny perturbations that they cause to the Earth's orbit, or to a conveniently passing comet. This left significant uncertainties until spacecraft visited them and experienced their gravitational pulls at close quarters.

The naming of moons

When moons began to be discovered there was no system for naming them. Galileo gave his four moons of Jupiter the collective name *Medicea Sidera* (the Medician Stars) after the family of his patron Cosimo de' Medici. He distinguished them individually by Roman numerals I, II, III, and IV, working outwards from Jupiter. In 1614 Simon Marius proposed the names Io, Europa, Ganymede, and Callisto, after lovers of the god Zeus (the Greek equivalent of Jupiter). Galileo, perhaps aggrieved by Marius's rival claim to discovery, would have nothing to do with those names. They did eventually become the officially recognized names that we use today, though a system of Roman numerals continues in parallel use (for newer discoveries as well). However, Galileo is honoured too, because as a group the four moons that he discovered (which are much larger than any of Jupiter's other moons) are called the 'Galilean moons'.

Cassini referred to the four moons of Saturn that he discovered as *Sidera Lodoicea* (the Louisian Stars) after his patron Louis XIV of France, alongside the larger *Luna Saturni* that Huygens had discovered. Other astronomers preferred a Galileo-style Roman numeral convention to identify individual moons. However, the practice of numbering outwards from the planet was a recipe for confusion, because new discoveries could change the order. For example, Huygen's *Luna Saturni* successively bore the numbers II, IV, and then V. By international consensus the numbering system was frozen after William Herschel's 1789 discoveries, so that the designations of individual bodies would never again be reshuffled. In 1847 William's son, John Herschel (1792–1871), suggested a coherent set of names for all seven of the then-known moons of Saturn, which we still use today. He called the largest one Titan after the collective name for the mythological siblings of Chronos (the Greek Saturn), and named the others after some of the individual Titans: Iapetus, Rhea, Tethys, Dione, Enceladus, and Mimas. When Lassell co-discovered the next moon of Saturn in 1848, he named it Hyperion, after another of the Titans in accordance with John Herschel's theme.

After he discovered the third and fourth moons of Uranus, Lassell invited John Herschel to name them along with his father's two previous discoveries. He chose Titania and Oberon (after the faerie queen and king in Shakespeare's *A Midsummer Night's Dream*), Ariel (after a sky spirit in Alexander Pope's *The Rape of the Lock* who also appears in Shakespeare's *The Tempest*), and Umbriel (a melancholy spirit in *The Rape of the Lock*).

Surprisingly, Lassell did not name the moon of Neptune that he discovered in 1846. Its present name, Triton (son of Poseidon, the Greek Neptune), was suggested as late as 1880.

The names of Mars' two tiny moons, Phobos and Deimos, were chosen by their discoverer based on a suggestion by Henry Madan (1838–1901), a science master at Eton school, from Book XV of *The Iliad* in which Ares (the Greek Mars) summons the twin brothers Fear (Phobos) and Terror (Deimos).

The names of subsequently discovered moons (there have been no more at Mars) have followed the theme that became established at each planet when the first names became generally accepted. Since its foundation in 1919, a body known as the International Astronomical Union (IAU) has been the arbiter of names (and their spelling) for objects in the Solar System, and also their surface features. A newly discovered moon is awarded a temporary designation (for example S/2011 J2 is the second satellite of Jupiter to have been discovered in 2011), and receives formal name only when its orbit has been well characterized. Names for Jupiter's moons are taken from lovers and descendants of Zeus/Jupiter. It is fortunate that mythology provides plenty to choose from, because Jupiter has fifty named moons and sixteen still awaiting formal naming.

Saturn has a similar number of known moons, of which fifty-three have been named. The first eighteen were named after Titans and their offspring. John Herschel's original theme was then broadened to include other giants from Greco-Roman, Gallic, Inuit, and Norse mythologies, allocated according to their orbits.

Except for Belinda, which is another name deriving from Pope's *The Rape of the Lock*, the moons of Uranus continue to receive names of (mostly female) Shakespearean characters. These range from the exotic Sycorax (a witch who was mother to Caliban in *The Tempest*) to the prosaic Margaret (a maidservant in *Much Ado About Nothing*). Neptune's moons are allocated names of mythological characters associated with Poseidon/Neptune.

How many moons?

It is hard to keep track of how many moons are known in our Solar System. Since the year 2000 there have been scores of discoveries. These were made mainly using ground-based telescopes and the Hubble Space Telescope, but NASA's Cassini spacecraft in orbit about Saturn found several additional small moons of that planet. The bodies known to possess moons include all the planets from Earth out to Neptune. It seems certain that neither of the inner planets, Venus and Mercury, can have a moon bigger than a kilometre or so in size, or it would have been discovered by telescopes or by orbiting spacecraft.

The 190 known moons of our Solar System's planets are listed in tables in the Appendix. There I quote 'mean radius' mainly because the smaller moons, especially those less than about 200 km in radius, can be markedly non-spherical in shape. For comparison, the Earth's mean radius is 6,371 km. I list density in thousands of kilograms per cubic metre. This is equivalent to tonnes per cubic metre, or grammes per cubic centimetre. For comparison, using the same units the density of water is 1.0.

Various icy bodies beyond Neptune have moons too. Among those, Pluto has the largest known family, and this is included in a table in the Appendix. Pluto-like bodies are substantial in size, and it is perhaps no great surprise that some of them should have moons. However, several asteroids, including one example less than a kilometre across, have been found to have moons too (which by definition are even smaller than the object that they orbit). Few people expected that, and it is a topic to which I will return in Chapter 7.

Does the Earth have more than one moon?

You may come upon claims on the Internet or in quizzes that the Earth has one or two small moons in addition to the Moon. These are misguided. Although there are very many artificial satellites in space, no small natural objects have been found in permanent orbit around the Earth. Any that there are must be less than about ten metres across or they would have already been discovered, and calculations suggest that such 'minimoons' could not have stable orbits.

However, there are small asteroids whose orbits cross that of the Earth in a way that enables them to become captured temporarily, during which interval they are indeed minimoons. An example is the five-metre-wide object 2006 RH_{120}. This was discovered telescopically in 2006, and the following year it made four loops round the Earth, each one different in shape. Its path was mostly well beyond the orbit of the Moon, but its closest approach brought it to about 70 per cent of the Moon's distance from us. It broke free after eleven months as a temporary satellite of the Earth, its orbit having been complex and unstable partly because it felt the Moon's gravitational pull as well as the Earth's. By 2017 it will be on the far side of the Sun, but will pass close again in 2028, offering another opportunity for temporary capture. What will happen can't be predicted, because irregular-shaped objects like this are small enough for solar radiation pressure to affect their trajectories. We can't estimate the size of the radiation-driven perturbations, because we know neither the object's shape nor its density.

Actually, 2006 RH_{120} may not be a natural object; it could be the third stage of one of the Apollo rockets used in NASA's 1969–72 Moon programme. Irrespective of this, some studies suggest that at any one time the Earth may be accompanied by up to a few dozen temporarily captured orbiters less than two metres in size that arrive, complete at least one loop, and are then lost. Larger examples are rarer, and it is probably only once in 100,000 years that something in the hundred-metre size range becomes temporarily captured in this way.

Another class of objects sometimes misreported as 'moons' of the Earth are asteroids whose average orbital period about the Sun is exactly the same as the Earth's. Such an asteroid has a path round the Sun that lies very close to the Earth's orbit, but the Earth's gravity exerts a strong influence on it, so that it cycles through a repeating pattern as follows. Imagine the asteroid following an orbit slightly inside the Earth's orbit. This means that the asteroid will be progressing round the Sun slightly faster than the Earth, and eventually it will catch up with it. As it draws near, the Earth's gravity pulls the asteroid into a larger orbit, which takes longer to complete so that the asteroid is now travelling more slowly round the Sun and lags behind the Earth. Eventually, after they have both made several orbits, the Earth almost catches up with the asteroid and pulls it into a smaller orbit. Now travelling faster than the Earth, the asteroid draws ahead and we are back to where we started.

From the perspective of the Earth, the asteroid's path resembles a horseshoe (with the Earth in the incomplete part of the ring). Because of that, these orbits are sometimes described as 'horseshoe orbits', but it is important to realize that the asteroid does not orbit the Earth and is always travelling in a forward direction around the Sun. At least three such asteroids are known, of which the largest, 2010 SO_{16}, has a mean radius of about 300 metres. A few other asteroids have less circular orbits with periods matching the Earth's, which leads them to migrate around a mean position either ahead of or behind the Earth, without migrating around a complete 'horseshoe'. The best known of those is 3753 Cruithne, which has a mean radius of about 2.5 km. This too orbits the Sun, not the Earth, and is not a moon. You can find more about the strange orbits of such 'quasi-moons' at a link in Further reading.

Can moons have moons?

Planets go round the Sun, and moons go round their planets, so a natural question is whether any moon could have a natural satellite of its own—a moon of a moon. In the case of our own Moon, it has had several temporary moons in the form of artificial satellites that humans have placed there in recent years. However, these have all been in unstable orbits leading them to crash on to the lunar surface after a few years. This is because the Moon's gravity is insufficient to dominate the region of space that surrounds it, because the much more massive Earth is too close.

The Moon has a stable orbit round the Earth because the Earth's gravity is strong enough to outcompete the Sun's gravity to a distance of about one million km, a volume of space that is described as the Earth's 'Hill sphere', after the American astronomer George William Hill (1838–1914) who defined the concept. The Moon's orbit lies well inside this, and so enjoys long-term stability. The Moon's own Hill sphere is 60,000 km across, but any object orbiting the Moon even within its Hill sphere experiences a sufficient pull from the Earth's gravity to cause its orbit to shrink over time. It turns out that the same is true for moons of other bodies too, so that orbits about moons anywhere in the Solar System are not stable in the long term. The duration over which an orbit about a moon could persist varies from years to many millions of years, according to a complex balance of forces, but is much shorter than the 4.5 billion-year age of the Solar System.

Thus there are no moons of moons, and if one were to be discovered it would almost certainly be a short-lived situation.

Chapter 2
The Moon

Now for some individual moons in detail. I will begin with the Moon, because this is the one about which most is known, and of course you will have seen it for yourself.

The Moon has been referred to by that name as far back as can be traced in Germanic languages (Old English *Mōna*, proto-Germanic *Mǣnōn*). In Latin it was *Luna* (whence the English adjective *lunar*, French *Lune*, Spanish/Italian *Luna*, and the identically pronounced Russian Луна). In ancient Greek the Moon was *Σελήνη* (*Selene*) from which we get the prefix used in terms such as 'selenographic coordinates' (the latitude/longitude system used for mapping the Moon). The term 'moon' was eventually transferred by analogy to any object orbiting another planet, though as we have seen the term satellite was invoked for them by Kepler much more quickly.

The Moon is a substantial body. Only four other moons are bigger (and also more massive, in the sense of having a greater mass): Jupiter's Io, Ganymede, and Callisto, and Saturn's Titan. If the Moon were to orbit the Sun independently there is no doubt that it would be ranked among the 'terrestrial planets', which are the Sun's four innermost planets. Table 1 shows both the equatorial and the polar radii of these bodies, because their rotation distorts their shapes so that they bulge at their equators and are flattened

Table 1 The Moon and the terrestrial planets compared

Name	Equatorial radius/km	Polar radius/km	Mass/ 10^{24} kg	Density/ 10^3 kg m^{-3}
The Moon	1,738.1	1,736.0	0.07342	3.344
The Earth	6,378.1	6,356.8	5.9726	5.514
Mercury	2,439.7	2,437.2	0.3301	5.427
Venus	6,051.8	6,051.8	4.8676	5.243
Mars	3,396.2	3,376.2	0.6417	3.933

towards their poles. This makes the equatorial radius greater than the polar radius. Venus is an exception: it spins nearly 250 times slower than the Earth and shows no measurable flattening.

Although not much smaller than Mercury, the Moon contains considerably less mass because its overall density is lower. The reason for this is that Mercury has a very large, dense iron-rich core surrounded by a relatively thin rocky mantle and crust, whereas the Moon is made almost entirely of rock. Its core, if it even has one, is no more than about 300 km in radius.

Mercury has a thick molten zone in the outer part of its core. This is circulating, and because it is an electrical conductor it acts like a dynamo and generates a magnetic field that encompasses the planet, partially shielding its surface from bombardment by the charged particles (cosmic rays) that stream out from the Sun. The Moon lacks such a magnetic field, and its surface is exposed to whatever the Sun throws at it.

Like Mercury, the Moon's own gravity is too slight for it to retain the gases necessary for an atmosphere, but there are some gaseous atoms above the surface. For example, sodium and potassium atoms are released by 'sputtering' when the surface is hit by the solar wind, helium is added directly by the solar wind, and argon escapes from the lunar interior where it is produced by radioactive decay of an isotope of potassium. The total 'atmospheric pressure' of these atoms is about 3×10^{-15} (3 million billionths!) of the Earth's surface atmospheric pressure. This is so low that atoms are more likely to escape into space than to bump into each other, making the whole of the Moon's atmosphere equivalent to the very tenuous outermost zone of the Earth's atmosphere, which is known as the 'exosphere'.

The virtual absence of an atmosphere leads to a very large day–night range of temperature at the Moon's surface, which swings from about 120°C near the equator at noon to around –150°C at night. These day–night variations do not penetrate deeply into the lunar soil or 'regolith', and at about a metre below the surface the temperature is believed to be a fairly constant –35°C. There are some craters near the poles whose floors are never illuminated by the Sun, and where the surface temperature is permanently below about –170°C.

Phases, orbit, and rotation

The Moon's appearance is familiar to most people. Even with the naked eye we can make out dark patches on its surface. The first telescopes enabled observers such as Galileo and Harriot (who did the better job) to map these dark patches and also make out smaller features such as craters and mountains. If you look with a pair of binoculars, you will probably be able to see more than they did.

Whenever you look, you will see the same hemisphere of the Moon as the one drawn by Harriot and Galileo, because the same side of the Moon always faces the Earth. What does vary is how much of it we can see lit by the Sun at any particular time, depending on where the Moon is in its 29.5-day orbit. When the Moon lies between the Earth and the Sun (which we call new Moon) we can't see it at all. It rarely comes *exactly* between the Earth and the Sun, causing a solar eclipse, because the Moon's orbit round the Earth is tilted at about 5° relative to the Earth's orbit round the Sun, with the result that the Moon usually passes (invisibly) either a little above or below the Sun rather than across its face.

A couple of days after new Moon, the Moon has drawn far enough away from the Sun in the sky for it to become visible as a thin crescent, which grows until half the Earth-facing hemisphere is lit. Somewhat confusingly, this is called 'first quarter', which refers to a quarter of the 29.5-day cycle being completed rather than to how much of the disc is illuminated. The visible illuminated area continues to grow (its shape is now referred to as 'gibbous'), until, nearly fifteen days after new Moon, the Moon is on the far side of the Earth from the Sun. The Earth-facing hemisphere is then fully lit, and we call this 'full Moon' (though if the Earth gets *exactly* in the way it stops direct sunlight reaching the Moon, and there is a lunar eclipse). As the Moon continues in its orbit, the illuminated fraction shrinks until only half the Earth-facing hemisphere is lit (third quarter), then the shape passes through a waning crescent until it disappears close to the date of the next new Moon.

If you think that seeing the same side all the time means that the Moon doesn't rotate, think again. To keep the same side facing the Earth, the Moon has to rotate exactly once per orbit. This is demonstrated in an animation to which I have put a link in the Further reading. Almost every known moon is in such a state of 'synchronous rotation', because of tidal drag that forces its rotation to keep pace with its orbit.

This comes about because the mutual gravitational attraction between two nearby bodies, such as a planet and its moon, distorts their shapes slightly. A tidal bulge (about a hundred metres in the case of the Moon) is raised in the middle of each facing hemisphere. This is because the nearside is closer to the other body than its centre is, and so experiences a slightly stronger gravitational pull. There is an equal bulge on the farside because the body's centre is pulled more strongly towards its neighbour than its farside is.

If a moon's axial rotation were faster than its orbital period, its shape would have to distort continuously as the tidal bulge migrated around the globe so as to stay lined up with the planet. This would use up energy until the moon's spin had been slowed down to match the orbital period. So although the Moon and most moons of other planets were probably spinning faster to begin with, their spin now matches their orbital periods.

Incidentally, although the Moon has a nearside and a farside, it doesn't have a permanent dark side. The 'dark side of the Moon' is an acceptable metaphor for the hidden side about which we once knew nothing. However, because the Moon rotates all sides see the Sun over the course of a single orbit—though of course the Sun can illuminate only half the Moon at any one time.

Although the Moon is orbiting the Earth, it is at the same time accompanying the Earth in its year-long journey round the Sun. The Earth's speed round the Sun is about 30 km/s, which is much faster than the Moon's orbital speed round the Earth: 1.022 km/s when the Moon is at its closest (a position known as perigee) and 1.076 km/s when at its furthest (apogee). Consequently, the Moon's path weaves to either side of the Earth's orbit, but it is always convex outwards from the Sun. It never backtracks.

The face of the Moon

We are the first generations to know what the Moon's farside looks like. This remained unknown until October 1959 when the Soviet probe Luna 3 was sent out beyond the Moon and beamed back the first blurry pictures. A much clearer modern view of the farside is included in Figure 2, which shows the Moon from four different directions. The nearside view shows what we can see from Earth

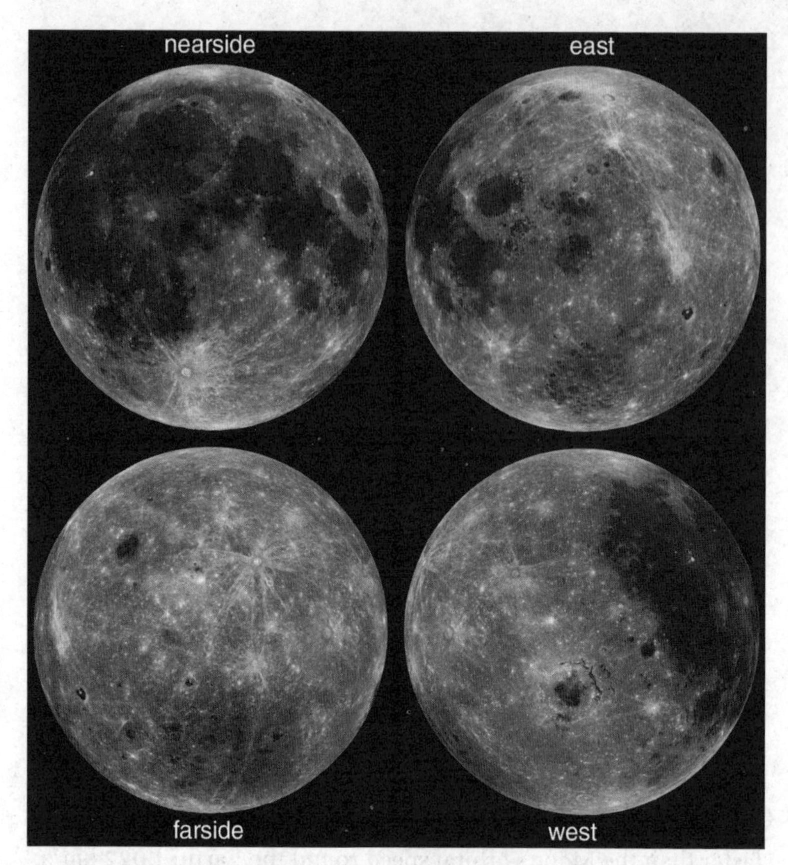

2. Four views of the Moon assembled from images acquired by NASA's Lunar Reconnaissance Orbiter. The nearside (centred at 0° longitude), the farside (centred at 180° longitude), and two intermediate views (east, centred at 90° longitude and west, centred at 270° longitude).

at full Moon. The farside view shows the opposite face of the Moon, seen when that side is fully illuminated by the Sun.

There is a striking contrast between the nearside and farside. About half the nearside is occupied by dark patches. In the 17th century, astronomers mistook these for seas (or at least the dry beds of former seas), and denoted them by the Latin word for sea, *mare*, of which the plural is *maria*. We now know that these areas have in fact been flooded by vast outpourings of volcanic lava resembling basalt. They have never been occupied by water, but *mare* and *maria* are still used today, both to refer to them in general and in their formal (IAU-approved) names. The correct pronunciations for *mare* and *maria* are 'MAH-ray' (not as in a female horse) and 'MAH-ria' (not as the girl's name).

On the farside only a few, relatively small, areas are occupied by maria. This is believed to be partially due to the crust being thicker on the farside, so that it has been harder for magma to reach the surface. It also suggests that the Moon has been in synchronous rotation with the same hemisphere facing the Earth since at least 3.8 billion years ago, when the majority of maria began to fill with lava.

The brighter surface outside of the maria is a different sort of rock. It is largely made of a kind of feldspar mineral called anorthite, giving the name 'anorthosite' to the rock. It represents the Moon's oldest crust, up to 4.5 billion years old. This terrain is referred to as the lunar highlands. Low-lying regions of 'highlands' on the nearside (most of them large basins excavated by impacts) were flooded by mare basalts, but this didn't happen to the same extent on the farside.

From Earth, we are able to see glimpses of alternate edges of the farside, because the Moon's orbit is not quite circular but an ellipse, which brings the Moon's centre to a distance of only 363,300 km from the Earth's centre at perigee but 405,500 km at apogee. Orbital speed is faster when closer and slower when further away, but the Moon's spin maintains a constant rate. As a result, around apogee the spin outpaces the orbital motion so we can see a little way round the average leading edge of the Moon, whereas the terrain close to the average trailing edge is rotated out of view. As perigee approaches, we see a little way beyond the average trailing edge instead. In total about 59 per cent of the Moon can be seen from Earth at one time or another, though we only ever get a foreshortened, highly oblique, view of the terrain near the edges.

This side-to-side wobbling of the Moon's face is a phenomenon called 'libration', and you can see it in action if you follow another of the links the Further reading. This also shows how the Moon looks slightly bigger (because it is closer) at perigee, and how the visible shape of the illuminated half of the Moon (the Moon's 'phases'), as seen from the Earth, changes as the Moon progresses round its orbit.

The images in Figure 2 lack any shadows, because the viewing geometry is directly down-Sun. This exaggerates the contrast in

3. The Moon as seen from Earth about three days eight hours before full Moon. The terminator is on the left. The illuminated edge of the visible disc (on the right) is referred to as the 'limb'.

the fraction of sunlight that different surfaces reflect (a property called 'albedo'), but topography is not apparent. Figure 3 shows a view of the gibbous Moon, in which the shadows are increasingly more obvious the closer you look to the terminator, which is the line dividing night from day.

In particular, many circular holes are apparent near the terminator thanks to shadows inside their rims. These are craters, whose origin I will discuss next, but first look at how their appearance differs between Figure 3 and the nearside view in Figure 2. The most obvious crater in the upper left of Figure 3 occurs on a dark background, because it lies in a region of mare. This is a 93 km diameter crater named Copernicus, thought to have formed about 0.8–1.1 billion years ago. It stands out strongly in the Figure 2 nearside view, because its floor penetrates through the low-albedo mare basalts into the buried high-albedo highland crust.

A prominent crater in the lower left of Figure 3 is in the middle of a bright patch in Figure 2. This is an 86 km diameter crater named Tycho. It has a brighter (higher albedo) floor than Copernicus because there is no basalt involved and it is all freshly pulverized highland crust. Some almost equally bright ejecta thrown out of the crater covers the surface surrounding it, and beyond this there are bright 'rays' radiating away, which are streaks of finely powdered ejecta. Apollo 17 landed about 2,000 km north-east of Tycho, but the astronauts collected fragments from one of Tycho's rays that crossed the landing site. Analyses carried out back on Earth found these to be about 100 million years old, showing that Tycho is much younger than Copernicus.

You can see several other rayed craters in the four images in Figure 2. Rays are prominent only when the Sun is close to overhead. They fade with age and are associated only with the youngest craters. Copernicus has rays, but they are less prominent than Tycho's.

Impact craters

Craters are circular rimmed depressions that dominate the Moon's surface. In fact they are abundant on the surfaces of most solid bodies in the Solar System, including almost every moon. They are rare only where ancient surface features have been erased by erosion, or buried by something such as lava flows or wind-blown dust.

In the case of the Earth, crust is recycled because of the action of plate tectonics, which crumples the edges of continents during collisions and pushes old ocean floor down into the mantle. This adds to the effects of general erosion and burial, so that the Earth has very few obvious Moon-like craters. Although nearly 200 have now been documented, few are spectacular, most are degraded, and none were well known when people began to notice and speculate about lunar craters.

It was Galileo who first referred to them as 'craters', using a Greek word meaning a drinking vessel. Their presence on the Moon undermined the notion that the Moon, being a heavenly body, should have a flawlessly smooth surface—though surely that was already hard to justify, given the blemishes represented by the obvious dark patches that can be seen even without a telescope, and which we now know as the maria.

The origin of lunar craters was a matter of controversy for some centuries. Most scientists regarded them as some kind of volcanic phenomenon, such as holes made by explosive eruptions or simply the scars left by vast bursting bubbles. The most common alternative theory—that they are the scars left by external objects hitting the Moon—was less popular for most of the time. Critics of that theory pointed to the fact that, with very few exceptions, the Moon's craters are circular in outline whereas impactors should arrive at all angles so that we would expect a large proportion of asymmetric or elongated craters. Others countered that lunar craters are very unlike what we see associated with most terrestrial volcanoes, so that a volcanic origin did not fit with the observations either.

It may seem obvious now that there are numerous bits of rock hurtling through the Solar System at high speed, but even as recently as 1808, US President Thomas Jefferson (by no means as ignorant of science as some of his successors) was publically sceptical. Of the results of inquiry by representatives of Yale University into a meteorite that was seen to fall and then collected in Connecticut, he is reputed to have declared that he 'would rather believe that two Yankee professors would lie than believe that stones fall from heaven'.

Evidence swung firmly in favour of an impact explanation for the Moon's craters in the 1960s. Three things helped with this. First, the American Gene Shoemaker (1928–97) studied the 1.2 km diameter Barringer Crater (also known as Meteor Crater) in Arizona, and found that the mineral quartz had been converted to denser forms of silica that form only under high pressure. He found identical mineralogic changes at underground nuclear test sites, and rightly concluded that the Barringer Crater had been excavated by high-pressure shock waves generated when a projectile struck the ground at tens of km per second (which is the sort of relative velocity expected when orbits intersect). Second, close-up images of the Moon's surface sent back by unmanned probes revealed that it has craters down to the smallest size visible. This could not be reconciled with any kind of volcanic phenomenon, whereas it posed no problem for impacts by objects hitting an airless body. Third, laboratory experiments resulted in convincing-looking craters with circular outlines when a 'hypervelocity' projectile was fired into a target in a vacuum chamber.

Figure 4 is a series of three nested and progressively higher-resolution (i.e. more detailed) views of lunar craters,

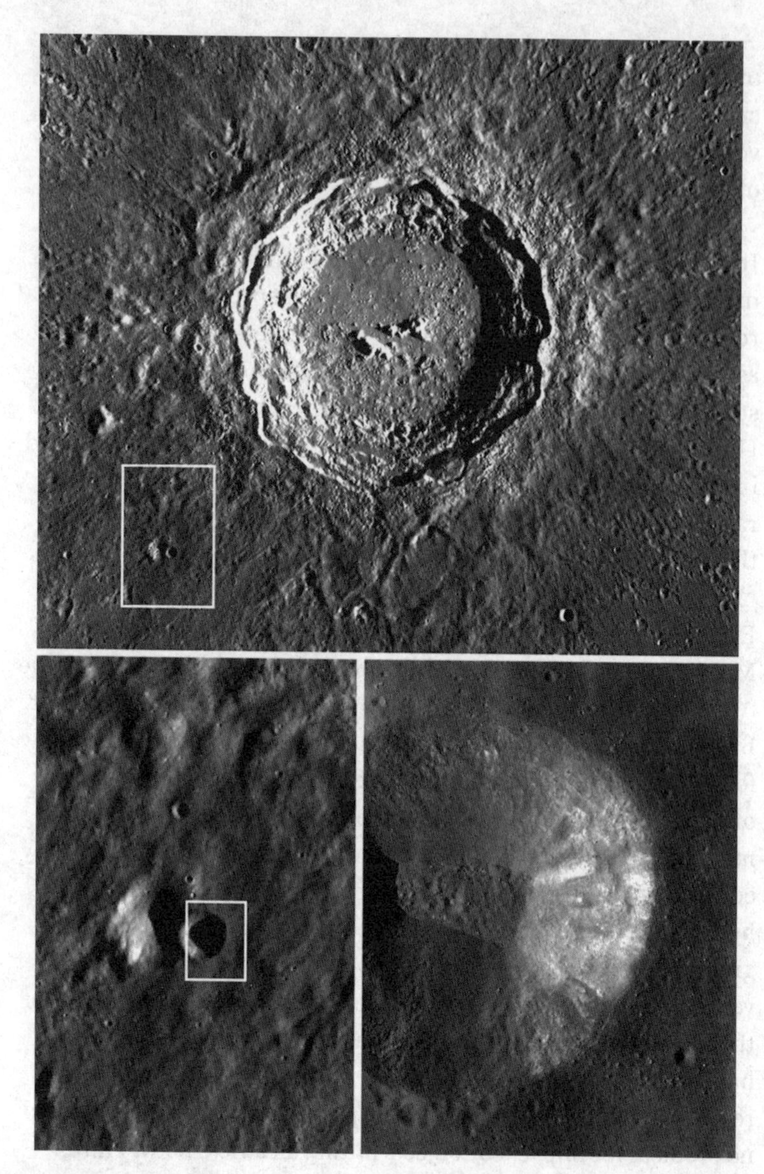

4. Copernicus crater and region seen by the Lunar Reconnaissance Orbiter at different resolutions. Top, Copernicus itself, seen at 250 metres per pixel. The image is about 200 km across, and the Sun was low in the east. The box outlines the higher-resolution view at lower left. Lower left, 20 km wide area seen at sixty-four metres per pixel. The box indicates the higher-resolution view to the right, the 7 km crater Copernicus B is immediately to the west of this box. Lower right, eight metres per pixel view of an unnamed 3.3 km wide crater,

beginning with Copernicus itself. Characteristics to note are that the floor of Copernicus is flat, except for a cluster of peaks near its centre. The crater floor is 2.5 km below the level of the plains surrounding the crater, and is bordered by a raised rim that is 1 to 1.5 km above the plains, giving a maximum drop from rim to floor of about 4 km. The central peaks rise about 1 km above the crater floor, so their summits are a long way below the crater rim and in fact do not rise even to the level of the surrounding plains.

The terrain outside Copernicus to a distance of more than one crater-radius is mantled by ejecta thrown out when the crater was formed. This buries any older craters, but the further away you look, the more smaller craters you can see. There are two near the middle of the medium-resolution view in Figure 4. The one to the left is about 7 km across, and is named 'Copernicus B'. It has a subdued shape as if its outlines have been smoothed by burial below Copernicus ejecta. An unnamed 3.5 km crater overlaps it, and has a much sharper appearance. This is younger, and was formed either by a 'secondary impact' when a large lump of Copernicus ejecta stuck the ground or at some later time by an unrelated small asteroid strike.

Neither Copernicus B nor its neighbour has peaks in its centre, and this illustrates an important observation: on the Moon central peaks are absent in craters less than about 10–20 km in size. Craters larger than that usually have a single central peak, unless they approach Copernicus in size when the single peak is replaced by a cluster of peaks. At even larger sizes—more than about 350 km across—crater structure becomes a double or even a multiple ring.

The highest-resolution view in Figure 4 was recorded with pixels only 8m across, and includes most of the younger crater that overlaps Copernicus B, which is about 300 metres deep. It was recorded with the Sun in the west and rather higher in the sky than for the two other views, so that the crater floor is free of shadows. At this level of detail you can see that the crater's inner slopes lack terraces of the kind so well seen in Copernicus. Outside the crater you can make out boulders as small as about twenty metres across and craters as small as about fifty metres.

There is no longer any doubt that the vast majority of craters on the Moon are a result of impacts. Asteroids, which are rocky or iron-rich bodies, typically strike the Moon at about 17 km/s. Comets, which are mainly icy bodies hurtling in from the outer Solar System, tend to strike faster at about 50 km/s.

As soon as an impactor hits the Moon's surface at speeds like this, shock waves radiate from the point of impact, which is why craters are circular and the angle of impact scarcely matters. The shockwaves melt and fragment both the target material and the impactor itself, and fling shattered and melted material outwards as ejecta. The first ejecta to be flung out travels the fastest, and goes furthest. During excavation, the crater gets wider rather than deeper. As the energy of the impact becomes expended the last ejecta barely manages to flop over the edge, contributing to the raised rim. The later ejecta falling close to the crater comes from deeper than the earlier ejecta, so any layering in the target material is inverted in the ejecta deposit.

Central peaks form, in sufficiently large craters, by a process called 'elastic rebound'. You can see similar rebound in high-speed videos of water droplets hitting the surface of a pond, except that in impact cratering the peak becomes frozen in place before it has chance to subside.

Craters end up thirty to a hundred times wider than the projectile that caused them, and the whole process is very fast. At typical impact speeds, to make a 93 km crater such as Copernicus would require a rocky asteroid about 5 km in diameter or an icy comet about 4 km in diameter, and would take about a couple of minutes to accomplish. The inner walls of craters as large as this tend to be unstable, and they subsequently collapse into a series of concentric terraces. To make a 7 km crater like Copernicus B would require a 200-metre asteroid or a hundred-metre comet, and would be all over in about thirty seconds.

Fortunately for us, because they would hit the Earth even more often than the Moon, impactors capable of making Copernicus-sized craters are now very rare. However, during the first billion years of the Solar System's history there was much more debris around, and some of it was big enough to make craters thousands of kilometres across (usually referred to as impact basins). There is no trace of these on the Earth any more, but many survive on the Moon.

On the nearside, these giant craters subsequently became flooded by lavas. Circular outlines attest to their origin, though some nearside basins have become sufficiently overfilled by lava that the flooded regions have merged. In the upper left of the nearside view in Figure 2, and also in Figure 3, you should be able to make out the 1,146 km diameter Mare Imbrium. This has an arc of mountains along its south-eastern edge, which is a relic of the basin rim. If you look at the west view in Figure 2, you will see the Orientale basin to the south of the equator. This has an inner ring about 500 km across partly filled by mare basalts within a 920 km outer ring that is mostly basalt free. The nearside maria are probably all double-ring, or multiple-ring structures like this, but have been so fully flooded that no trace of the inner rings is visible at the surface.

The Orientale basin straddles the boundary between nearside and farside. The largest mare entirely within the farside is Mare Moscoviense, which is in the upper left of the farside view in Figure 2. This fills the inner ring of a double-ring basin nearly 500 km in diameter, and spills out into part of the outer ring. There is a much larger basin on the farside that has little or no mare basalt within it. This is the 2,500 km diameter South Pole–Aitken basin, which is the largest and oldest basin preserved on the Moon. It has excavated to a depth of 13 km and shows up as an area of medium albedo (darker than highlands, but lighter than mare) stretching north from the south pole in the Figure 2 farside view.

Craters and dating

The Moon's craters have turned out to be very useful for working out the sequence of events that has affected the Moon. Working this out just requires common sense. For example, a crater that is superimposed on another must be younger, and (more usefully) the mutual relationship between dispersed ejecta from one crater and another that lies amidst the ejecta shows which of the two is youngest. There are also diagnostic relationships between craters and mare basalts, as Figure 5 illustrates.

There is a lot to see in this image. For now I will discuss just the relationships between the craters and the mare basalts that fill the whole of this field of view. The crater Beketov and those labelled D, E, and L are simple bowl-shaped craters, and there can be no doubt that these are younger than the mare basalt surface. However, what about Jansen? This is 24 km in diameter, easily big enough to have a central peak. However, it has a flat floor only about 150 metres below its rim. The explanation is that the Jansen crater was already there when the mare basalts flooded this area. The basalt lava flows overtopped the rim of the crater, flooding and burying its central peak. Subsequent degassing and thermal contraction caused the lava surface to subside, which allowed the rim of the buried crater (but not its more deeply buried central peak) to re-express itself on the topography. Crater Y punches 550 metres into Jansen's lava-flooded floor, and was made by an impact that happened much later.

Crater R, slightly larger than Jansen, is scarcely discernible at all. You can make out its circular outline within an otherwise normal lava surface. This is what is known as a 'ghost crater', denoting an old crater so completely flooded that it hardly shows up at all.

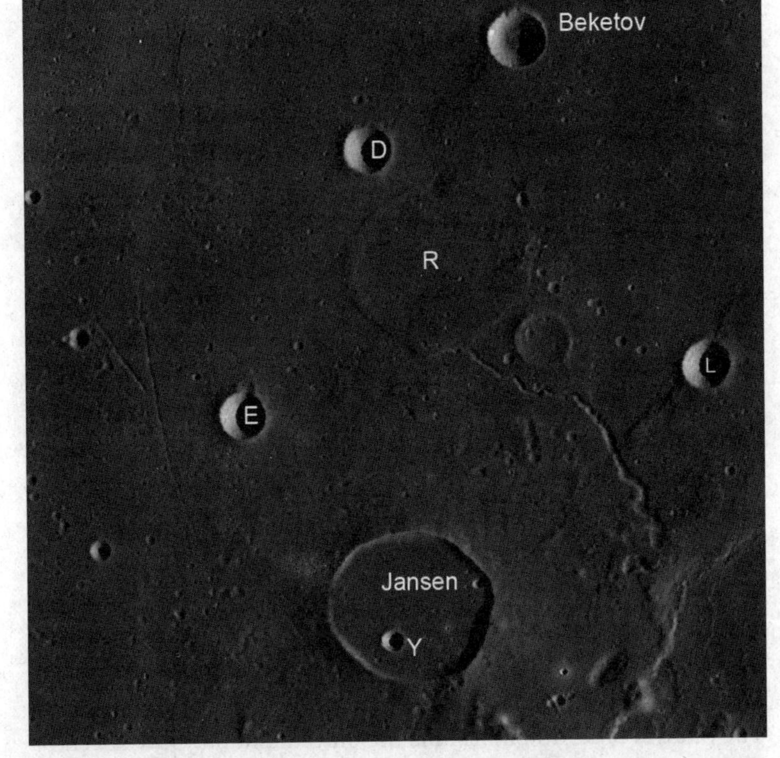

5. A 110 km wide region in the north of Mare Tranquilitatis, including Jansen crater. The letters are the IAU designations of Jansen's 'satellite craters'.

There is a smaller (unlettered) ghost crater just outside it, to the south-east.

The sequence of events that we can deduce from these simple observations is as follows. First, there was an impact that formed the large basin that is now occupied by Mare Tranquilitatis. Next, there were impacts into the floor of that basin, to make craters such as Jansen itself, Jansen R, and Jansen R's unnamed smaller neighbour. Then came the flooding of the basin by mare basalt lava flows, which presumably completely hid some craters but was not sufficient to fully obscure the craters now visible as 'ghosts'.

The new surface was subsequently struck by the impactors that made Beketov, Jansen D, E, L, and Y, and numerous smaller craters.

The lunar cratering timescale works on the basis that the longer a surface has been around, the more craters will have formed on it. This can be tested using mutual age relationships such as we have just looked at. It has been calibrated into an absolute timescale using samples of rocks and minerals brought back from the Moon that have been dated in the laboratory by measuring the accumulated products of radioactive decay. This shows that the rate of crater formation was intense during the interval 4.1–3.8 billion years ago, an epoch known as the late heavy bombardment. No lunar surfaces have survived from before this time, so we are not sure what was happening even earlier. The term 'late' denotes late in terms of the formation of the Solar System. In fact it was early in lunar history, the Moon and the Earth being about 4.5 billion years old.

Thirty known basins formed during the late heavy bombardment. The South Pole–Aitken basin is the oldest, and may have been made about 4.1–4.0 billion years ago. The Imbrium basin was formed about 3.8 billion years ago, and the Orientale basin (the youngest of its kind) is about 3.7 billion years old.

In all cases, it was several hundred million years after basin formation that the most extensive mare basalts were erupted. Most had been emplaced by about three billion years ago, but eruptions continued in some areas until about one billion years ago. Over the past 3.5 billion years the rate of cratering seems to have been fairly constant and much lower than during the late heavy bombardment, though it is impossible to rule out short flurries of bombardment.

Cratering has not ceased. As already mentioned, material in one of Tycho's rays (and hence Tycho itself) has been dated at a hundred million years old. Smaller, brand new craters have been imaged by lunar orbiters, including an eighteen-metre crater whose formation was marked by a brief flash recorded by ground-based telescopes on 18 March 2013. The impactor that made this crater would have been about a metre across, and would not have survived to make a crater if it had encountered the Earth, because it would have been vaporized by friction in our atmosphere.

Names of craters and maria

The naming of craters on the Moon is now under the control of the IAU. Most are named after deceased scientists and polar explorers. A few more recently named craters honour cosmonauts and astronauts who lost their lives. Names of maria are mostly Latin terms describing weather conditions (for example, Mare Imbrium means 'Sea of Rains'). One exception is Mare Moscoviense (Moscow Sea) that was named when discovered on the first farside images from Luna 3. Another is Mare Orientale (Eastern Sea), a name that might seem perverse for a feature on the central longitude of the Moon's *western* hemisphere. Part of it can be seen from Earth during favourable libration, and the name is inherited from a time when Earth-bound astronomers used directions in the Earth's sky rather than thinking about coordinates from the perspective of someone on the Moon's surface.

Given the history of the naming of Jupiter's moons, you will probably not be surprised to learn that early observers put forward conflicting schemes for lunar nomenclature. The present-day scheme derives from that of an Italian Jesuit named Giovanni Riccioli (1598–1671), published in 1651. The convention of applying letters to identify smaller craters within or around a larger named crater (such as Copernicius B in Figure 4 and Jansen D, E, L, R, and Y in Figure 5) was devised by the German Johann Mädler (1794–1874). Note that these letters recognize proximity alone. Jansen D, E, L, and Y must have formed long after Jansen itself, so their origin is certainly not related to Jansen and probably not to each other.

Mare basalts and regolith

The mare basalts appear to have formed when parts of the Moon's mantle grew hot enough for 'partial melting' to begin, enabling some melt to be sweated out, leaving a solid residue in the mantle. The magma that reached the surface was similar to basalt in composition. This has lower viscosity than most molten rock, and was able to spread far across the surface before solidifying.

The early stages of eruption may have been violent, with molten lava flung hundreds of metres skywards before falling to the ground. The main fissures from which the lavas were erupted cannot be seen, because they were filled in and covered by later more gentle effusions, but some buried fissures filled by dense solidified lava were identified in gravity mapping by NASA's GRAIL mission in 2012 (Table 2). Some later lava flows carved channels into the surface of earlier flows, and there is a nice example in Figure 5 beginning about 30 km from the eastern rim of Jansen and winding towards Jansen R. West of Jansen E you may be able to make out a couple of straight, narrow fissures, which overlie dyke intrusions or are cracks that formed as the lava cooled and contracted.

Although the Moon is rightly described as a rocky body, there is actually very little solid bedrock exposed at the surface. Impacts of all sizes have distributed a layer of fragments (estimated to be about ten metres thick on the highlands and five metres thick on the mare surfaces, which are younger) that makes up the lunar soil, more properly called 'regolith'. Most of the regolith is made of fragments less than a tenth of a millimetre in size, though impacts

Table 2 Highlights of lunar exploration

Name	Country	Date	Achievements
Luna 1	USSR	4 Jan. 1959	Fly-by, no pictures
Luna 2	USSR	13 Sept. 1959	Impact on to Moon
Luna 3	USSR	6 Oct. 1959	Fly-by; first farside pictures
Ranger 7	USA	31 July 1964	Impactor
Luna 9	USSR	3–6 Feb. 1966	Lander; first pictures from the surface
Luna 10	USSR	3 Apr.–30 May 1966	First lunar orbiter
Surveyor 1	USA	2 June 1966–7 Jan. 1967	Lander
Lunar Orbiter 1	USA	14 June 1966–29 Oct. 1967	Orbiter
Apollo 8	USA	24–7 Dec. 1968	First manned orbiter
Apollo 11	USA	20–1 July 1969	First manned landing; 21.5 kg of samples
Apollo 12, 14–17	USA	Nov. 1969–Dec. 1972	Manned landings; 360 kg of samples
Luna 16	USSR	20–4 Sept. 1970	First robotic sample return; 0.1 kg

(*continued*)

Table 2 Continued

Name	Country	Date	Achievements
Lunokhod 1	USSR	17 Nov. 1970–14 Sept. 1971	First lunar rover; 11.5 km traverse
Luna 20, 24	USSR	Feb. 1972, Aug. 1976	Robotic sample returns 175 kg
Lunokhod 2	USSR	15 Jan.–11 May 1973	Lunar rover; 40 km traverse
Hiten	Japan	Mar. 1990–Apr. 1993	Orbiter/impactor
Clementine	USA	Feb.–June 1994	Orbiter
Lunar Prospector	USA	Jan. 1998–July 1999	Orbiter/impactor
SMART-1	Europe	Nov. 2004–Sept. 2005	Orbiter
SELENE (Kaguya)	Japan	Oct. 2007–June 2009	Orbiter/impactor
Chang'e 1	China	Nov. 2007–Mar. 2009	Orbiter/impactor
Chandrayaan-1	India	Nov. 2008–Aug. 2009	Orbiter and impactor
Lunar Reconnaissance Orbiter	USA	June 2009–	Orbiter
LCROSS	USA	9 Oct. 2009	Impactor
GRAIL	USA	Jan.–Dec. 2012	Gravity mapping from orbit
Chang'e 3	China	14 Dec. 2013–	Orbiter, lander, and rover (Yutu)

6. Apollo 15 astronaut James Irwin and the Lunar Roving Vehicle on the brink of Hadley Rille (a 1.5 km wide, 300-metre-deep lava channel) in July 1971. Note the footprints in the regolith in the foreground.

big enough to penetrate into bedrock fling out some boulder-sized pieces too, like the example in the left of Figure 6.

The dominantly dusty nature of the regolith means that footprints are sharp, and indeed the paths trodden by Apollo astronauts in the years 1969–72 can be seen today in super-high-resolution images taken from lunar orbit.

Lunar exploration

The Moon was the prime target for exploration until the US won the 'space race' with its landing of Apollo 11 in July 1969. After the ending of the Apollo programme and the Soviet unmanned sample return efforts (three landers that brought back just short of a third of a kilogram of regolith between them) in the 1970s, there was an interval of nearly twenty years before exploration picked up again, with a new generation of unmanned orbiters equipped for specific scientific observations. More nations joined in too, as can be seen in Table 2. Looking to the future, India plans an orbiter/lander/rover package named Chandrayaan-2 for 2016–17 and China plans a sample return mission named Chang'e 5 in 2017. An ambitious crowd-funded project called Lunar Mission One was begun in the UK in 2014, intended to land a probe to drill into the floor of the South Pole–Aitken basin in 2024. It is not clear when humans will return to the Moon, but if I had to place a bet it would be on a Chinese mission in about 2020.

'And don't forget to bring back some rock!' was the gag in a newspaper cartoon depicting someone waving goodbye to the Apollo 11 crew before launch. Bringing samples of the Moon back to Earth, where they can be subjected to detailed and precise analysis, was the most important scientific goal of the project. I have already mentioned radioactive dating. Other analyses include determining what minerals make up lunar rock and whether the abundances of different elements in those minerals differs from what we would find in rocks from Earth. For elements such as oxygen that have more than one stable isotope, the relative abundances of these isotopes can be used as a 'fingerprint' to see whether the Earth and Moon were made from the same source material. Even something as basic as examining a rock with a microscope to study its texture can tell us things about the rock's origin and history that can't be deduced from orbit.

The six Apollo landings collected a total of 382 kg of samples, and the three successful robotic Luna missions brought back a further 0.32 kg. All of these were from the nearside, so some important regions remain unsampled. The South Pole–Aitken basin is perhaps the most compelling target for future sample return.

However, there is more Moon rock available on Earth than what has been deliberately brought back. Impacts on to the Moon can throw a fraction of their ejecta out with sufficient velocity that it escapes the Moon and may eventually land on the Earth as a meteorite. Lunar meteorites were first recognized in 1982. They are different to the more common types of meteorites that come from asteroids, though if we didn't have previously known samples of the Moon to compare them with their origin might still be in doubt. About 48 kg of lunar rock has now been collected as meteorites. About half of this is likely to come from the farside, but of course we cannot identify the specific location where any example originated.

Some lunar rocks are breccias, made of pieces of basalt and/or highland rock smashed apart by impacts but also welded together by the heat generated by impacts. Other samples are lumps of a single basalt or highland rock type. Regolith contains fragmented or pulverized relics of all these, and also glassy beads about 0.1 mm in size that appear to be frozen droplets of basaltic spray from explosive eruptions.

Most mare basalts in and around Mare Imbrium show geochemical fingerprints of having been derived from a source region relatively rich in potassium (K), rare earth elements (REEs), and phosphorous (P), known as KREEP. It is thought that this records an anomalous patch in the nearside mantle in which heat from radioactive decay of potassium (and also thorium, which has been mapped from orbit by detecting its emitted gamma rays) was a key cause of the partial melting that led to the eruption of the mare basalts.

An early recognized characteristic of lunar samples shown by the analytical techniques available at the time was a lack of water bound up inside minerals, nor any chemical alteration of minerals. Mineral crystals billions of years old look as fresh as if they grew only last year, whereas on the Earth the damp environment would have led to chemical alteration penetrating into minerals along fractures and cleavage planes.

The Moon is a very dry place. In addition to scarcity of water, it has a much lower inventory of 'volatile elements', such as sodium.

On the other hand, the relative abundances of the three stable isotopes of oxygen in lunar rocks is an almost perfect match to what we find in the Earth, which suggests that the two bodies formed from the same source material. However, if that were the case, why does the Moon not have an iron core to match the Earth's core? Small core and depletion in volatiles on the one hand, and similar oxygen isotopes on the other, are apparent contradictions that have to be reconciled when trying to explain how the Moon formed in the first place.

The Moon's origin

In 1879 George Darwin (1845–1912), second son of his more famous father Charles, proposed that the Moon formed by fission—splitting off from a previously more rapidly spinning Earth. This could explain the Moon's lack of a core and oxygen isotope match to the Earth, but not its depletion in volatiles. A co-accretion model, in which the Earth and Moon grew side by side while the Solar System was forming, falls down on the lack of a core as well as mismatched volatiles. Capture of the Moon by the Earth after independent formation would be dynamically very difficult, and would require the oxygen isotope match to be a fluke.

The Moon's origin is still a matter of debate. A theory that it was formed by a 'giant impact' on to the early Earth was developed in the mid-1980s and has become widely accepted, because it could explain the matches and mismatches between the two bodies. The latter stages in the growth of terrestrial planets are probably a series of a few collisions (giant impacts) between bodies of roughly similar size (known as planetary embryos), rather than growth of a larger body by numerous impacts of much smaller bodies. Giant impacts usually result in a merger of the two planetary embryos, leaving a larger embryo expected to be largely molten because of the heat generated by the impact. This melting would make it easy for iron (which is dense) to sink inwards to form a core; a process described as differentiation.

According to the giant impact hypothesis of the Moon's origin, the final giant impact experienced by the proto-Earth was when a differentiated Mars-sized planetary embryo struck it a glancing blow. Rather than leading to full merger, the collision ejected the impactor's mantle and some of the Earth's mantle into space, whereas the impactor's core ploughed inwards and merged with the Earth's core. The Moon then grew in orbit round the Earth, from a mixture of the impactor's mantle and the target's mantle.

Energy converted to heat as the Moon grew would have melted it, allowing what little free iron there was to sink to form a tiny core. More importantly, as the Moon's global magma ocean began to cool, the first crystals to form would have been anorthite, the very mineral discovered to make up most of the lunar highlands. Anorthite has a relatively low density, and the crystals would tend to rise. Probably they would need to clump together to form larger masses before their buoyancy could overcome the magma's viscosity, but then they would rise until they breeched the surface, coalescing to form the lunar highland crust.

Radioactive dating has determined that the oldest lunar crust samples were formed about 4.35 billion years ago, about 200 million years after the birth of the Solar System, but the Moon-forming impact could have occurred as long ago as 4.5 billion years.

Water and individual volatile elements would have been preferentially lost to space before the debris from the giant impact was able to coalesce into the Moon, so that the giant impact hypothesis is the most credible story. The giant impactor has even been given a name—Theia, after the mythological Selene's mother. A recent variant of the hypothesis, invoked to explain some of the differences between nearside and farside, is that the debris originally formed two moons, which merged in a relatively slow mutual collision a few tens of millions of years after the giant impact.

Water on the Moon

The Moon is not so entirely bone dry as it seemed to be, based on the original interpretation of the Apollo data. This century, traces of water have been found within lunar samples in a type of mineral called apatite. Apatite's crystalline structure causes it to incorporate any available water molecules, and also any large negative ions such as those formed by fluorine and chlorine. There is an ongoing debate about how to use apatite to interpret the water content of the Moon's mantle. It could be anything from ten parts per million to one part per thousand. We aren't sure what the average water content of the Earth's mantle is either, but it is almost certainly higher than the Moon's, perhaps about 1 per cent, and so the Moon's surface rock and its interior are still regarded as much drier than the Earth.

However, there is a second, unrelated, reservoir of water on the Moon. This occurs as ice inside the cold, –170°C, craters near the poles whose floors are never illuminated by the Sun. Evidence for this built up slowly. In 1994 the Clementine orbiter showed that radio waves bounce from those craters in a way consistent with ice. Five years later, the neutron spectrometer on the Lunar Prospector orbiter demonstrated concentrations of hydrogen that could most reasonably be interpreted as residing in water (H_2O). The clincher came in 2009, when the Centaur rocket that had delivered the Lunar Reconnaissance Orbiter was crashed into the permanently shadowed floor of a crater named Cabeus near the south pole. A probe called LCROSS (Lunar Crater Observation and Sensing Satellite) followed six minutes behind. Before it too crashed, it was able to analyse the ejecta flung up by the Centaur crash, showing the floor of the Cabeus to consist of about 6 per cent water by mass.

There is a fairly simple explanation for the Moon's polar ice, and it has nothing to do with any water from *inside* the Moon that has been there since the Moon's birth. When a comet hits the Moon and makes a crater the comet's ice is vaporized. The water molecules are added to the Moon's exosphere. If a molecule hits a hot surface, it will bounce, and will eventually be lost to space. However, if it hits a cold surface it will stick there. If it hits a permanently cold place, it will stick there indefinitely. In this way permanently shadowed crater floors act as 'cold traps' where ice accumulates molecule by molecule, and continues to do so today. The same process builds up ice inside Mercury's polar craters too.

Chapter 3
The Moon's influence on us

The Moon's presence in the sky was a spur to our initial move into space, but it has long pervaded human culture in many other ways, including providing themes for music and song (both good and bad) and influencing language. For example, 'waxing and waning' refers to the Moon's illuminated face growing or shrinking as it goes through its phases.

The Moon has had almost as much influence as the Sun in the way we keep time. It takes the Moon 27.3 days to complete an orbit round the Earth, but because the Earth moves about a twelfth of the way round the Sun in that time, it takes 29.5 days for the Sun–Earth–Moon relationship to repeat. This 29.5-day cycle of phases is the origin of the months that have been used in virtually every known human calendar. In the English and Germanic languages the words for Moon and month even share the same root. Dividing a lunar month into four (for example, the time between first quarter and full Moon) is probably the origin of our seven-day week.

Western cultures, and for official purposes the whole world, use a solar calendar. This defines the start of the year by a fixed position in the Earth's orbit round the Sun. To keep things neat there are twelve months in a 365- or 366-day year, of which all except February are slightly longer than a lunar month. However, the

Islamic calendar is lunar, and counts time in years of 354 or 355 days (twelve 29.5-day lunar months), which are thus shorter than the Earth's orbital period round the Sun. Traditional Chinese and related oriental calendars are lunisolar, using lunar months but defining the new year as the second (sometimes the third) new Moon after the winter solstice, which means that years are of different length but average out to be equal to the Earth's orbital period.

The saying 'once in a blue Moon' originally referred to an infrequent or even impossible event. Recently, calendricists have formalized the expression 'a blue Moon' to refer to the occasion when a second full Moon occurs within the same calendar month, which averages out to about seven times in every nineteen years.

Although everyone should be used to seeing the Moon in the sky, there is a surprisingly common mistake. Visualize a crescent Moon, or better still ask an unsuspecting friend or a child to draw one. More than likely they will come up with a crescent resembling a letter ☾, rather than its mirror image ☽. They will do this even if they live north of the tropics, where the Moon looks like a ☾ when it is a waning crescent, which can be seen only in the dawn sky. I have no idea why this is. Most of us see the crescent Moon far more often in the evening sky, when from the northern hemisphere its shape is a ☽.

Seen from the southern hemisphere, objects in the sky look 'upside down'. There, the waxing Moon in the evening sky looks like a ☾, whereas the waning Moon in the morning sky is the other way round. The ☾ and ☽ shapes that I have described are angled, with the illuminated arc tilted downwards (towards the Sun, near or below the horizon), and the tilt becomes greater the closer you are to the tropics. From a viewpoint near the equator, if you see a crescent Moon in a dark sky it will always look like a ◡ (never ◠), because the illuminated edge has to be facing towards the Sun, which is below the horizon, either having recently set or being about to rise.

Ocean tides

Tides in the oceans are caused by the gravitational pull of the Moon and the Sun on the ocean water, which can respond much more freely than the solid Earth. The Moon is far less massive than the Sun, but this is more than compensated by its relative closeness to the Earth, with the result that the tidal force exerted by the Moon is just over twice as strong as the solar tide.

When the Moon and Sun are lined up at new Moon there are two high tides and two low tides in the day. These will occur close to noon and midnight, except where coastlines complicate the movement of water (for example round the British Isles). An important point is that there are two tides per day, not one, because the solid Earth is pulled towards the Moon and Sun while being pulled away from the ocean water on the nightside. The tidal effect is exactly the same at full Moon, because although the Moon and Sun are pulling in opposite directions it is the *difference* in the strength of their pulls on opposite faces of the Earth that determines the tide. At such times, because the solar and lunar tides add together, the tidal range (the amount by which the tide rises and falls) is greatest. This situation is referred to as a 'spring tide', a term that refers to rapid rise and is nothing to do with the season. Conversely, when the Sun and Moon are at right angles to each other in the sky (at first quarter and last quarter), the weaker solar tide acting six hours out of phase with the lunar tide decreases the tidal range. This situation is called a 'neap tide'.

These effects are summarized in diagrammatic form in Figure 7. The solid bodies of the Earth and Moon are distorted by the same forces, but by a much smaller amount than the oceans.

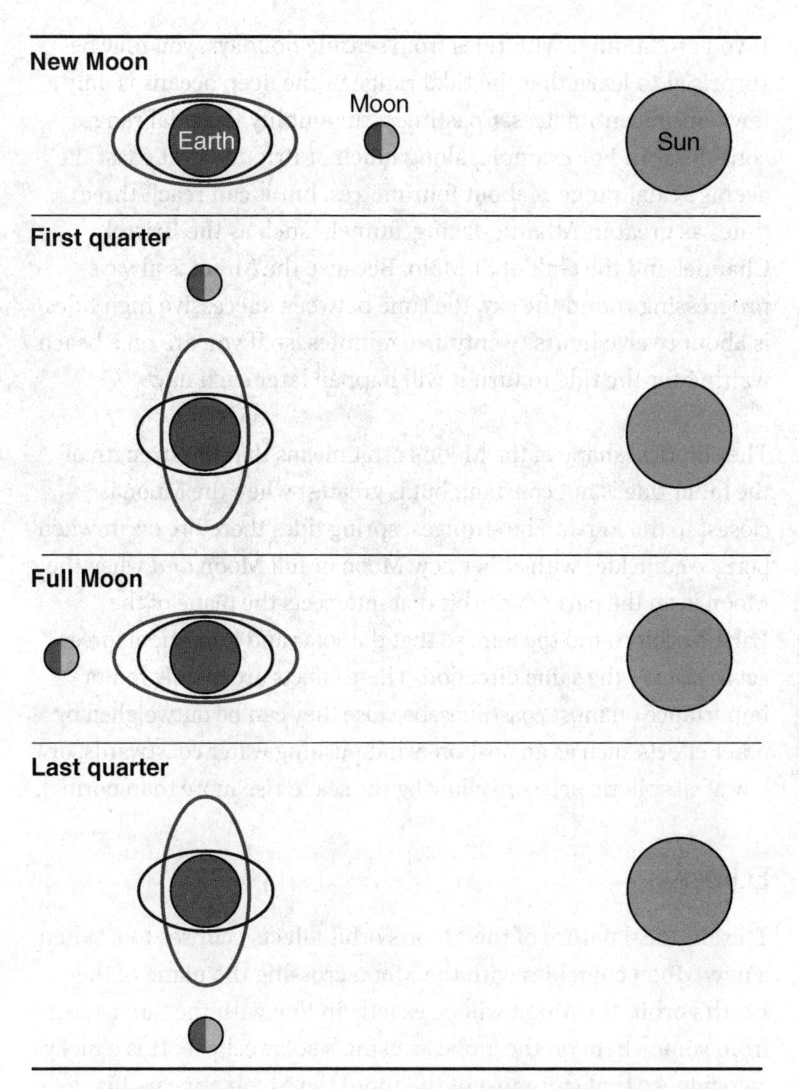

7. Sun, Earth, and Moon (not to scale) at four successive times to illustrate ocean tides. The ellipses round the Earth show the strength of the tide-producing force from the Moon (the larger ellipse) and from the Sun (the smaller ellipse). The effects add at new and full Moon to produce the highest tidal range (spring tides), whereas at first and last quarter the maximum solar and lunar tides are different by 90°, resulting in a smaller tidal range (neap tides).

If you are familiar with tides from seaside holidays, you may be surprised to learn that the tidal range in the deep oceans is only a few tens of centimetres. Coastlines can amplify the tidal range considerably. For example, along much of Britain's east coast the average tidal range is about four metres, but it can reach three times as great in Atlantic-facing 'funnels' such as the Bristol Channel and the Gulf of St Malo. Because the Moon is always progressing round the sky, the time between successive high tides is about twelve hours twenty-five minutes, so if you are on a beach waiting for the tide to turn it will happen later each day.

The elliptical shape of the Moon's orbit means that the strength of the lunar tide is not constant, but is greatest when the Moon is closest to the Earth. The strongest spring tides therefore occur when perigee coincides with either new Moon or full Moon *and* when the Moon is on the part of its orbit that intersects the plane of the Earth's orbit round the Sun, so that the solar and lunar tidal forces act in *exactly* the same direction. These effects are of only minor importance on most coastlines, because they can be outweighed by other effects such as an onshore wind pushing water coastwards, or low atmospheric pressure allowing the sea to rise more than normal.

Eclipses

The elliptical nature of the Moon's orbit affects eclipses too. When a new Moon coincides with the Moon crossing the plane of the Earth's orbit, the Moon will be exactly in line with the Sun as seen from somewhere on the globe, causing a solar eclipse. It is a lucky coincidence that (for most of the time) the Moon appears big enough to blot out the whole solar disc for up to seven minutes, allowing the Sun's otherwise invisible outer atmosphere (the corona) to be seen in all its glory. However, when a solar eclipse happens with the Moon at apogee, the Moon is too far away to completely hide the Sun, so at the time of exact alignment an annular eclipse is seen, with a bright ring of the Sun's disc surrounding the part obscured by the unlit Moon.

Because the Moon is relatively close to the Earth, parallax means that you have to be in the right place to see a total eclipse. The path of the Moon's shadow across the globe is only about 250 km wide even at perigee. Outside of this path, you can see only a partial eclipse where the Moon hides only part of the Sun.

On the other hand, when the Moon is exactly opposite the Sun in the sky, the Earth casts a shadow big enough to cover the whole Moon. This is called a lunar eclipse, and can be seen from anywhere where the Moon is above the horizon. Although the eclipsed Moon is in the Earth's shadow, it remains visible because it is lit by reddened sunlight that has been bent through the Earth's atmosphere. Looking at the Moon from the Earth, you see a dull red disc. Looking from the Moon, you would see a red 'sunset glow' rimming the Earth.

Orbital recession and day length

Just as the Earth raises tidal bulges on the Moon, so the Moon's pull raises tidal bulges in the solid shape of the Earth, in addition to the ocean tides. Because the Moon is so much less massive, its effect on the Earth is much less than the Earth's effect on the Moon. Thus the Moon has not managed to slow the Earth's rotation to match the Moon's orbital period (Pluto and its largest moon Charon are a rare example where this has happened).

Measurements show that the Earth's rotation is slowing by about 1.6 milliseconds per century, which is mostly a result of tidal drag from the Moon. The same forces are slowing the Moon's orbital speed (and, at the same time, its rotation). Slower orbital speed requires it to be further from the Earth, and it is receding at a rate of 3.8 cm per year. Recession of this sort can be calculated, but it has also been demonstrated by determining the gradually lengthening travel time of light between the Earth and Moon by directing laser beams on to retroreflectors left on the lunar surface by Apollos 11, 14, and 15 and the Soviet Lunokhods.

If the Moon's present-day rate of recession had been constant, the Moon would have been too close to the Earth to be stable between one and two billion years ago, so it seems that recession must be faster now than in the past. We have no way of measuring how close the Moon was in the distant past, but there are indications of how fast the Earth was spinning. Some varieties of coral show daily growth lines when examined under a microscope. In corals that are 370 million years old, these lines come in annual cycles of 400, showing that there were 400 days per year. The Earth's orbital period is unlikely to have changed by much, so we can conclude that the day length was shorter (about twenty-two hours) in order to fit 400 days into a year.

The Moon's influence on human behaviour

The very word 'lunacy' is derived from the Latin *lunaticus*, meaning a person affected by a supposed Moon-induced madness. However, whenever a sufficiently careful test has been done there has been found to be no correlation between the phase of the Moon and mental illness, or indeed any kind of human behaviour including crime, suicide, or birth rate. There is no physical reason to expect any influence. For example, the tidal forces experienced by the human body are vanishingly small.

It has been widely believed in most cultures that the Moon has an influence on human fertility, probably because of the similar length of the lunar month and the human menstrual cycle. However, this seems to be a coincidence, and no connection between the two has been shown to exist.

There is, however, a modern example of the Moon influencing human behaviour, and this is over the matter of so-called 'supermoons'. The Moon's angular size in the sky is about 14 per cent greater at perigee than at apogee, so that when perigee coincides with full Moon, the full Moon is bigger than average. The difference is too slight to notice unless you actually take some measurements, but despite this, over the past few years whenever full Moon falls close to perigee there has been a fuss on social media websites and in the conventional news media too, claiming that the phenomenon will be spectacular or even that it may cause a natural disaster of some sort.

The term 'supermoon' was coined in the 1980s by an astrologer, to denote full Moon occurring when the Moon is in the 10 per cent of its orbit that is closest to perigee. This catchy term has caught the public imagination and has replaced the more cumbersome phrase 'perigee full Moon'. However, it is misleading. A supermoon is only slightly bigger than average, and it is not even a particularly rare phenomenon (it is much more common than a 'blue Moon'). About one full Moon in every ten qualifies as a supermoon, and in fact there were three in succession in July, August, and September 2014. Contrast the commonness and trivial difference in apparent size of a supermoon with the amazing powers possessed by the fictional 'Superman', or with a 'supervolcano eruption'. The latter denotes a catastrophic eruption that produces at least ten times more ash than any volcano has managed in the past 1,000 years, and is an event that happens somewhere on Earth only once in about every 50,000 years on average. No correlation has been found, or is likely, between supermoons and events such as eruptions and earthquakes.

Although its angular size in the sky is bigger at perigee, the Moon's surface isn't actually any brighter then. Surface brightness depends on the Moon's distance from the Sun, not its distance from the Earth. The dominant factor is the eccentricity of the Earth's orbit, which leads to a roughly five million-kilometre difference in Earth–Sun (and Moon–Sun) distance between nearest (in early January) and furthest (in early July). This swamps the 40,000 km variation in Earth–Moon distance, and leads to a 7 per cent range in the Moon's surface brightness. However, because the Moon's angular size in the sky is greater when it is closest, for a given Earth–Sun distance full Moon at perigee does provide about 30 per cent more moonlight than when a full Moon occurs at apogee.

The human eye adapts to different levels of brightness. Our pupils dilate or contract to allow a comfortable amount of light on to our retinas, but our memories do not store this information. Thus, without using instruments we have no way to make a quantitative comparison between one full Moon and another several weeks or months later, and anyone who just looks at a supermoon and claims that they can tell that it is bigger or that the moonlight is brighter than normal is fooling themselves.

In contrast to the hyped-up and almost entirely imagined supermoon phenomenon, there is an unrelated optical illusion that can genuinely make the Moon look big. Whenever we see the Moon low in the sky (the closer to the horizon the better), the Moon seems to be large. It isn't really any bigger, and if you measure it the size doesn't change. This is an effect called 'the Moon illusion', and seems to be a result of the way your brain processes information. When the Moon is high in the sky, it seems small—isolated and adrift in a sea of back. However, when it is near the horizon there are distant but familiar objects with which to relate it, such as trees and rooftops, beyond which the Moon looms large.

The Moon and life on Earth

The Moon affects the behaviour of wildlife in numerous ways. Nocturnal land animals tend to vary their activity with the phase of the Moon, according to whether the extra light around full Moon is helpful or harmful to their success as predators or safety as prey. Many marine species use the Moon as a clock to trigger mass spawning, not because the light level matters, but because the success of this reproductive strategy depends on synchronization. Some turtles wait for spring tides so they can come ashore easily and lay their eggs somewhere that will stay dry until their young hatch.

The Moon may have influenced life on Earth in more fundamental ways than merely affecting feeding and breeding habits. It has been suggested that without the Moon to cause tides, it would have been much harder for life to migrate from the sea on to land. That's probably wrong, because the Sun on its own would cause twice-daily tides with about half the range that occurs in the average combined lunar and solar tides. This would offer plenty of scope for marine organisms to find themselves temporarily stranded by a falling tide.

Another influence of the Moon is that its presence might stabilize the tilt of the Earth's axis. Currently this is 23.4°. The axis points to a direction in space that changes only very slowly, and this tilt is responsible for the seasons as the Earth progresses round its orbit. Calculations suggest that over the past five million years the tilt has varied by about 2.5° with a period of about 41,000 years. This affects the climate, as can be seen in the fossil record. Studies in the 1990s suggested that if the Earth had no Moon, our axial tilt would experience much wilder fluctuations, ranging from zero to nearly 90°. This would have led to major extremes of climate, even worse than those experienced by Mars, which has only tiny moons and where axial tilt is currently varying between about 5° and 45°.

Such extremes of climate could have made it impossible for advanced life to establish itself on land, in which case if the Moon did not exist then we wouldn't either. However, more recent studies have contested the significance of the Moon, and suggested that the Earth's axial tilt could stay within narrow bounds even without the Moon's influence.

Moonbases and lunar resources

When humans eventually return to the Moon, it will probably lead to a more sustained lunar presence than the few days achieved by each Apollo landing. There are many good scientific reasons for sending trained astronauts (preferably geologists) if the goal is to explore the Moon. Humans can use their time much more efficiently than remote-controlled rovers. In 1973 it took the Soviet Lunokhod 2 nearly four months to travel further than the Apollo 17 astronauts managed in three excursions totalling twenty-two hours.

Apart from learning more about the Moon, being on the Moon would enable other science too. On the farside, the Moon shields you from radio interference from the Earth, making a farside radio telescope an attractive prospect for astronomers. Bizarrely, the Moon may also be the best place to find evidence of conditions on the early Earth. Just as impacts on to the Moon fling off ejecta that can be collected as meteorites on the Earth, so impacts on to the Earth must throw out ejecta that could be collected on the Moon. A piece of the Earth's surface that was ejected on to the Moon three billion years ago would be a wonderful find, because it would be in pristine condition compared to any weathered relict still on Earth. We would hope to find tiny but analysable samples of the early Earth's atmosphere trapped as bubbles within shocked glass, and maybe microfossils documenting the first stages of life on Earth.

It is debatable whether the Moon has any resources that it would make economic sense to bring back to Earth. The cost of getting to the Moon in order to fetch anything is very high. Because the Moon is so dry, it would be surprising to find ore minerals concentrated by the action of aqueous fluids circulating through the crust like we do on Earth, but there might be tempting concentrations of platinum-group elements hidden within craters made by the impact of metallic asteroids.

One commodity that it might make sense to export from the Moon to the Earth is an isotope of helium called helium-3. This is rare on Earth, but samples from the Moon show about ten parts per billion of helium-3 in lunar regolith, believed to have been implanted there by the solar wind. Helium-3 and a heavy isotope of hydrogen ('deuterium', which can be extracted from seawater) are the fuel required for nuclear fusion reactors, a proposed 'clean' source of electric power. The technological viability of fusion power has yet to be demonstrated, so all this is rather speculative. However, if a market for helium-3 ever does emerge on Earth, it would become commercially viable to strip-mine the surface regolith of the Moon, heating it to drive off, and then collect, the helium-3 gas.

Using the resources of the Moon *on* the Moon is a different matter. Water extracted from a permanently shadowed crater could make a nearby Moonbase much cheaper to run. Solar panels on a hilltop beside a polar crater could gather uninterrupted solar power that could be used to heat the regolith and to form it into building blocks, or to drive more complex processes to extract oxygen and metals, or to make glass fibre. Even in the case of habitat units brought from Earth, it could make sense to simply pile regolith over them to offer protection from solar storms.

Who owns the Moon?

There is no agreed legal framework to establish ownership of the Moon or any part of it. A United Nations 'Agreement Governing the Activities of States on the Moon and Other Celestial Bodies', instigated in 1979 and known for short as the Moon Treaty, declares that the Moon should be used for the benefit of all states and all peoples of the international community. It also expresses a desire to prevent the Moon from becoming a source of international conflict. These are fine sentiments, but the Treaty is toothless. Of the nations capable of flying independently to the Moon, only India has signed (though not ratified) the Treaty.

The Treaty also forbids ownership of extraterrestrial property by any organization or person, unless that organization is international and governmental. If this ever became effective it would be a major hindrance to the commercially driven exploration of the Moon and the rest of the Solar System. A more pragmatic approach might be to accept that if someone invests in exploration and can find a market for a resource, then they should be allowed to sell it to whomever wants to buy it—whether on-planet or off.

A claim, of sorts, to small areas of the Moon for non-commercial motives is enshrined in the Apollo Lunar Legacy Bill, which went before a US Congressional Committee in 2013. This seeks to give protected status to the wheeltracks, footprints, and hardware at the six Apollo landing sites. This seems reasonable to me. These are part of humanity's communal heritage, and it would be awful if some 'space pirate' could with impunity mess up the sites, or hack away bits of the hardware to sell privately on Earth.

8. Private property on the Moon? Top, the Lunar 21 lander, imaged from the Lunokhod 2 rover in January 1973. Bottom left, Luna 21 and right, Lunokhod 2: imaged in 2010 and 2011 by the Lunar Reconnaissance Orbiter.

It's a grey area, but most legislators would agree that nations or organizations that have landed an object on the Moon still own that object, though not necessarily the ground on which it stands. In 1993 the Lavochkin Association, a Russian aerospace company, sold the Luna 21 lander (Figure 8) and the Lunokhod 2 rover, both of which it had built, at a Sotheby's auction. These fetched a price of $68,500 and the purchaser was the British/American Richard Garriott (1961–), a video game developer and entrepreneur, who later made a self-funded visit to the International Space Station.

Chapter 4
The moons of giant planets

The giant planets Jupiter, Saturn, Uranus, and Neptune each have an extensive entourage of moons. Naturally, the first moons to be discovered were the largest, and are often called the 'regular satellites', but they are only part of the story. Not all the moons of giant planets can be neatly pigeonholed, but the overall situation is as follows.

Closest to the planet are small inner moonlets, mostly less than a few tens of kilometres in radius and irregular in shape. They are closely associated with the planet's ring system and their orbits are circular, lie in the planet's equatorial plane, and have radii less than about three times that of the planet itself.

Next are large regular satellites exceeding about 200 km in radius, which is large enough for their own gravity to have pulled them into near-spherical shapes, a condition described as 'hydrostatic equilibrium'. Their orbits are only slightly less circular than those of the inner moonlets, and have radii up to twenty or thirty times that of the planet. These too lie pretty close to the plane of the planet's equator.

Finally there are the irregular satellites, mostly less than a few tens of kilometres in radius. The term refers both to their irregularity in shape and to their orbits, which can be strongly elliptical and are usually considerably inclined relative to the planet's equator. They extend to about 400 times the radius of Jupiter and Saturn, over 800 times the radius of Uranus, and nearly 2,000 times the radius of Neptune.

Inner moonlets and all regular satellites except for Neptune's Triton travel round their orbits in the same direction that their planet rotates, which is described as 'prograde' motion. Most irregular satellites, as well as having inclined orbits, travel round their orbits in the direction opposite to their planet's spin. This is described as 'retrograde' motion, and has implications for these moons' origins.

Ice

Generally speaking, large moons are not rocky bodies. In the 1950s, telescopes fitted with spectrometers to measure the characteristics of reflected sunlight and trained on the larger moons of the giant planets began to reveal the presence of frozen water on most surfaces. This was not really surprising, because these bodies are a long way from the Sun, and mean surface temperatures are about –160°C for the moons of Jupiter, –180°C at Saturn, –200°C at Uranus, and –235°C at Neptune. At such extremely low temperatures, ice is mechanically very strong and behaves like rock; it can sustain craters, cliffs, and mountains without flowing like a glacier would on Earth.

Just as importantly, the temperature was also very low when these moons were forming. Jupiter, five times further from the Sun than is the Earth, lies beyond the 'ice line'. Temperatures here were low enough to allow water to condense directly into ice from the gas cloud surrounding the young Sun. Bodies that formed beyond the ice line generally contain more ice than rock, because in the cloud of gas and dust from which the Solar System formed the elements required to make water (hydrogen and oxygen) were more abundant than the key ingredients of rock (silicon and various metallic elements plus oxygen). Where hydrogen could form solid compounds, it did so, so rock dominates only inside the ice line.

Carbon and nitrogen are common elements too, and these went to make up other varieties of ice that condensed further from the Sun. Possibly at Saturn, and certainly at Uranus and beyond, the ice is not just water but is mixed with frozen ammonia (NH_3), methane (CH_4), carbon monoxide (CO), and (at Neptune) even frozen nitrogen (N_2). In the giant planets much of this, especially water, occurs as ice in their interiors below a thick gassy envelope mainly of hydrogen and helium, but the moons have too little gravity to have collected a lot of gas, so ices dominate.

The abundance of ice explains the low densities of most moons. The tables in the Appendix list densities of between 1,000 and 2,000 kg per cubic metre for most regular satellites of the giant planets. A rocky body should have a density of more than 3,000 kg per cubic metre, whereas water ice has a density of 1,000 kg per cubic metre (other ices are even less dense). Thus the lower its bulk density, the more ice and the less rock a moon contains.

Missions to moons

There would be much less to say were it not for space probes that have visited the giant planets and their moons. Exploration began with fly-bys (missions that flew past the planet) but has moved on to the stage of orbital tours in the case of Jupiter and Saturn, which have each had a mission that orbited the planet for several years and that was able to make repeated close fly-bys of at least the regular satellites. Close fly-bys of moons enable detailed imaging, and usually take the probe close enough to see how the moon affects the strong magnetic field surrounding the planet and to detect whether the moon also has its own magnetic field. The size of the slight deflection to a probe's trajectory as it passes close to a moon enables the moon's mass to be determined. Knowing the moon's size, it is then easy to work out its density.

The story begins with NASA's Pioneer 10 that flew past Jupiter in December 1973, and Pioneer 11 that flew past Jupiter in December 1974 and then Saturn in September 1979. These were concerned mostly with the planets' atmospheres and magnetic fields, and collected little data about their moons.

It was NASA's two Voyager probes that really opened our eyes to the moons. Voyager 1 flew through the Jupiter system in March 1979 and through the Saturn system in November 1980. In August 2012 it became the first space probe to cross the heliopause, where the solar wind fails, and to enter interstellar space. Voyager 2 made fly-bys of all four giant planets: Jupiter in July 1979, Saturn in August 1981, Uranus in January 1986, and Neptune in August 1989. It remains the only probe to have visited Uranus or Neptune.

NASA's Galileo mission went into orbit around Jupiter in December 1995. After dropping an entry probe into Jupiter it toured the moons for eight years until it ran out of manoeuvring propellant and was allowed to crash into the planet. There was a serious early set back because its main communications antenna, a parabolic dish, failed to deploy. This meant that data had to be transmitted using the backup 'low-gain' antenna, reducing the total number of images that could be collected, but in-flight programming and data compression techniques rescued much of the science.

The joint NASA–European Space Agency (ESA) mission Cassini–Huygens arrived at Saturn in June 2004. It released the Huygens lander that parachuted to the surface of Titan in January 2005, while the Cassini orbiter began a long and complex orbital tour that is scheduled to end with entry into Saturn's atmosphere in 2017.

Cassini flew past Jupiter in December 2000 on its way to Saturn, and for several days was able to complement the Galileo orbiter's observations of volcanic eruptions on Io. More images of these spectacular events were provided by NASA's New Horizons mission, which made a close pass by Jupiter in February 2007 on its way towards Pluto, which it flew past in July 2015.

Jupiter's regular satellites

Jupiter's four Galilean moons are the archetypal regular satellites. I consider them here as a family, reserving individual treatment for Chapter 5. They are shown together in Figure 9, cut away to reveal their internal structures. These were deduced mainly from clues to their internal density distribution achieved by Galileo fly-bys together with measurements by Voyager and Galileo of the interplay between each moon and Jupiter's magnetic field. The latter shows that Jupiter's magnetic field induces a field within Europa and Callisto, most likely achieved by electrical conduction in a salty internal ocean. Ganymede has a fairly strong magnetic field of its own, which may be generated by convection currents acting like a dynamo in a liquid iron sulfide outer zone of its core, as happens inside Mercury and the Earth. Io's magnetic field has been less well characterized, and we cannot be certain whether it results from motion in a fluid core or is an induced field with a relatively shallow source. Three are differentiated bodies, in which the denser material has been able to segregate inwards to form a core, but Callisto lacks a strong internal density gradient, showing that it is only weakly differentiated.

Io is the densest moon in the Solar System and is the only regular satellite to lack surface ice. It can be thought of as a larger (and more active) version of our own Moon. It has a rocky surface, stained yellow and red by sulfur compounds distributed by ongoing volcanic eruptions.

Europa is a smaller (and less active) version of Io, buried by water, which is solid near the surface (ice) and liquid at depth where it forms a global ocean. Europa is nearly as dense as the Moon, and

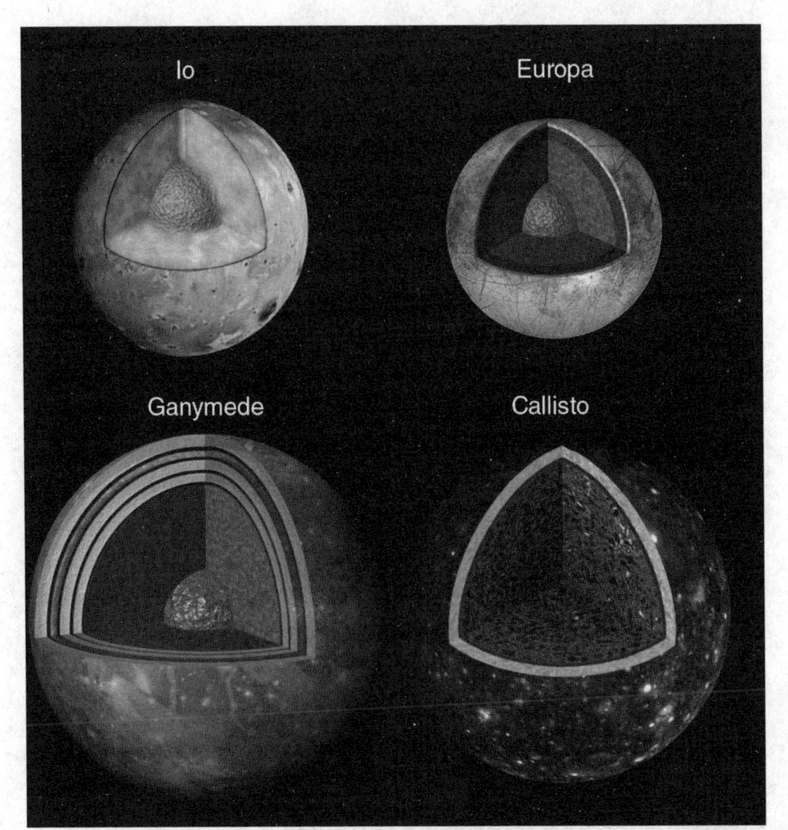

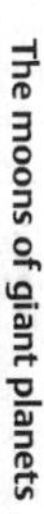

9. Cutaway views to show the inferred internal structures of Jupiter's regular satellites, shown to scale. Io, Europa, and Ganymede are shown with an iron-rich core, surrounded by rock. The outer layers of Europa, Ganymede, and Callisto are ice, and the dark layers inside are liquid water. Callisto's interior is shown as an undifferentiated mixture of rock and ice.

this shows that the shell composed of water (ice plus liquid) is only about 100 km thick. Europa's surface ice is able to fracture and migrate relative to the interior, but neither Europa nor any other moon shows behaviour closely similar to the formation and migration of tectonic plates as on Earth.

Ganymede is the most massive moon in the Solar System and is the only differentiated Galilean moon in which the internal pressure is sufficient to compact H_2O-ice into denser crystalline structures. Its surface ice is the normal sort, known as Ice I, but at greater depths the calculated pressure is enough to compress it into phases known as Ice III, then Ice V, and finally Ice VI. Figure 9 shows a complex internal structure proposed for Ganymede in 2014. According to this model, each ice phase has an underlying liquid layer, giving the appearance of a multi-layered sandwich. If these liquid layers exist, they are probably quite strong brines, containing salts dissolved out of the rocky interior that keep them liquid at temperatures too cold for pure water to melt.

The density of the Galilean moons decreases outwards from Jupiter. This is a strong clue to their origin, and suggests that they grew from a cloud of debris around the young Jupiter (in much the same way as the planets grew around the Sun), and that the heat radiated by Jupiter was sufficient to starve Io, and to partially starve Europa, of the water that was available to moons further out. Debris around Jupiter would have shared Jupiter's rotation, which would naturally result in these having moons prograde orbits close to the planet's equatorial plane, as indeed we see.

Orbital resonance and tidal heating

When the Voyager missions were being planned, the regular moons were expected to be fairly dull places. It was argued that, being largely icy and relatively small, the amount of heat-producing radioactive elements contained in any interior rock would be too little for there to have been any internally driven activity that could have affected their surfaces during the past three or four billion years. They would therefore be 'dead' worlds, heavily scarred by impact craters like the lunar highlands. Rock or ice, it doesn't matter: impacts cause craters that look much the same.

However, it turns out that Io has a surface so young that no impact craters at all have been seen there, and Europa has very few. There are plenty of impact craters on Ganymede and Callisto, some of which you should be able to make out on Figure 9, but Ganymede has tracts of paler, younger terrain cutting across its surface. So, as well as an outward trend of decreasing density, the Galilean moons have an outward trend of increasing surface age.

The nearer moons are not actually any younger, but they have been resurfaced more recently by geological activity. The explanation lies in their orbits. Tidal friction has long since slowed down their spin, so they have synchronous rotation matching their orbital periods. It has also made their orbits become much more circular than the Moon's, because they are orbiting a much more massive planet with stronger gravity.

In an elliptical orbit, libration would displace a moon's tidal bulges to and fro about their mean position. The distortion of the interior to allow this to happen must add heat by means of internal friction, and would encourage the interior to become differentiated, if the even more powerful tidal heating before its rotation became synchronous had not already done so. If you want to experience the efficacy of internal frictional heating, try bending a wire coat hanger to and fro, and then (carefully!) touch the bent part to your lip.

If Jupiter had only one moon, the planet's pull would have taken less than a hundred million years to force the moon's orbit into an exactly circular shape, whereupon there would be no more libration and no tidal heating. However, as you know Jupiter has four regular satellites. While Voyager 1 was speeding towards Jupiter, the American Stan Peale (1937–2015) and colleagues were computing the effect that membership of a family of moons can have on tidal heating.

The orbital periods of Europa and Ganymede are exactly twice and four times that of Io, so that for every four circuits of Jupiter made by Io, Europa completes exactly two and Ganymede exactly one. This is a situation described as 'orbital resonance', and has been brought about by mutual gravitational interactions between the moons. This means that Io passes Europa at exactly the same point in its orbit every time, and Europa passes Ganymede at exactly the same point in its orbit. The repeated slight gravitational tug between moons has prevented their orbits from becoming *exactly* circular. If you saw them drawn on paper they would look like circles, but there is enough departure from circularity for the orbital speed to vary so that the tidal bulges migrate slightly to and fro, and swell and contract in height, heating their interiors in doing so.

Peale and his colleagues calculated the amount of heating that this process should produce inside Io. Having made the best assumptions they could about Io's internal structure and strength, they published their findings in the 2 March 1979 issue of the journal *Science*, writing 'dissipation of tidal energy in Jupiter's satellite Io is likely to have melted a major fraction of the mass. Consequences of a largely molten interior may be evident in pictures of Io's surface returned by Voyager 1.' This is the most remarkable example of timing in planetary science known to me; images recorded by Voyager 1 six days later revealed erupting volcanoes on Io, and a total lack of surviving impact craters.

Orbital resonance is a complex situation. While in a state of resonance, the amount of forced eccentricity (and hence the rate of tidal heating) can wax and wane, and also moons can drift in an out of resonance over millions of years. Present-day tidal heating accounts for Europa's young surface, and past episodes of tidal heating can be invoked for Ganymede and several regular satellites of other giant planets.

Other regular satellites

The regular satellite families of the other giant planets lack the clear outward trend of decreasing density seen at Jupiter, and, with the exception of Saturn's Titan, the individual moons are smaller. However, the orbital characteristics at Saturn and Uranus are similar; almost circular orbits in their planet's equatorial plane, even at Uranus where some catastrophe long ago tipped the whole system on to its side (Uranus' rotation axis is tilted at 97.9° relative to its orbit round the Sun). At present there is orbital resonance between some of Saturn's moons but none among Uranus' moons, though some surfaces bear signs of tidal heating suggesting that resonances occurred in the past.

Saturn has seven regular satellites (see Appendix). Working outwards these are Mimas, Enceladus, Tethys, Dione, Rhea, Titan, and Iapetus. Mimas is in 2:1 resonance with Tethys, though this seems to have produced no effective tidal heating. Enceladus is in 2:1 resonance with Dione, and this must power the eruption plumes that Cassini discovered near Enceladus' south pole that are discussed in Chapter 5.

Uranus has five regular satellites: Miranda, Ariel, Umbriel, Titania, and Oberon. Of these, Ariel and Titania are crossed by large fractures best attributed to ancient tidal heating. Miranda has a complex surface history, possibly the result of a previous 3:1 orbital resonance with Umbriel. It is slightly non-spherical (its radii are 240, 234.2, and 232.9 km) and some might regard it as Uranus' outermost inner moonlet.

These all probably formed around their planets in a similar manner to that proposed for Jupiter's Galilean moons, but the same cannot be said for Neptune's largest moon, Triton (not to be confused with the similarly named Titan at Saturn). This has an inclined and retrograde orbit. There is no known way in which Triton could have formed alongside Neptune and ended up in such an orbit, so it was probably captured by Neptune from elsewhere.

Most likely Triton began as a member of the 'Kuiper belt', a family of icy bodies, from 1,500 km downwards in size, that orbit the

Sun near Neptune's orbit and beyond. If this is correct, Triton's originally independent orbit round the Sun must have, on one occasion, brought it close enough to Neptune to become captured. Capture is very difficult to achieve, because there is too much momentum to dispose of. Usually the approaching body would just swing past the planet, or (rarely) collide with it. However, if the incoming object is actually double (a primary object and a moon), one can be captured and the other can be flung away faster than it arrived, with the total momentum of the system being conserved.

Such double objects can in fact be seen today in the Kuiper belt. Pluto and its large moon Charon are a prime example. Triton is slightly bigger than Pluto, but we do not know whether or not it was the larger or smaller component of the supposed double object that arrived at Neptune. Whether double or single, Triton's arrival and capture would have scattered any pre-existing regular satellites, which must have been lost to the Kuiper belt or destroyed in mutual collisions.

Trojan moons

When a smaller body orbits a larger one, there are two stable points where an even smaller body can reside or about which it can oscillate. These occur 60° ahead and 60° behind the orbiting body, sharing the same orbit about the larger body. Technically these are called Lagrangian points, and are designated as L_4 (ahead) and L_5 (trailing). However, they are often referred to as the 'Trojan points' because there are groups of asteroids named after characters from the Trojan war that share Jupiter's orbit round the Sun, clustered around its L_4 and L_5 points.

Saturn has four small moons in leading and trailing Trojan relationships with two of its regular satellites, Tethys and Dione. Calypso, Telesto, and Helene, which are between 10 and 20 km in radius, were discovered telescopically in the 1980s and 1990s. The

10. The Solar System's four known Trojan moons, imaged by Cassini and shown at approximately correct relative scale. Helene has a mean radius of 17 km. Polydeuces has never been seen in sufficient close up to reveal surface details. Helene and Polydeuces are Trojan co-orbitals of Dione, whereas Telesto and Calypso are co-orbitals of Tethys.

smallest, Polydeuces, which is less than 3 km across, was discovered on images taken by Cassini in 2004.

Images from Cassini (Figure 10) have revealed some surprising aspects to the surfaces of these Trojan moons. They have rather few impact craters, and so may be relatively young (this could still mean they are more than a billion years old). The figure shows the side of Helene that faces away from Saturn and its rings; the Saturn-facing side has more craters. Calypso's surface is the most reflective in the entire Solar System, but this is not apparent in Figure 10 as the images have been processed to show each surface equally well. This may be a result of Calypso sweeping up ice crystals erupted from Enceladus. Both Calypso and Helene show curious gullies on their surfaces. There is no likely liquid that would be stable under their surface conditions. Some kind of dry avalanche process may be responsible, but it is a mystery how this would work on a body whose surface gravity is only 0.02 per cent as strong as the Earth's.

Saturn is the only planet known to have Trojan moons, but this may be a result of observational bias. The Galileo orbiter didn't search for them at Jupiter, which could have small ones like Polydeuces. Voyager 2 didn't have chance to look for any at Uranus and Neptune, where even Calypso-sized Trojans would be hard to spot telescopically from Earth.

Irregular satellites

There are more known irregular satellites than any other class of moons. Jupiter has fifty-nine. The inner seven are in prograde orbits up to 238 Jupiter radii (Jupiter's equatorial radius, 71,492 km) in size, and the others are all in retrograde orbits extending out to 400 Jupiter radii. This is a consequence of the differential long-term stability of orbits relative to the size of a planet's Hill sphere (the range out to which the planet's gravity outcompetes the Sun's gravity). Prograde orbits are stable over billions of years out to only about half the Hill sphere radius, whereas retrograde orbits can be stable out to about two-thirds of the Hill sphere radius. Jupiter's Hill sphere is about 740 Jupiter radii in size.

Jupiter's largest and third-closest irregular satellite, Himalia, is 85 km in radius and was discovered as long ago as 1904, but most have been found by dedicated telescope surveys since the year 2000. The smallest are only about 1 km in radius. Even Himalia is only seven pixels across in the best Galileo image, so little is known of any of them.

Three other prograde irregular moons have orbits similar to Himalia (at about 160 Jupiter radii), and they all reflect sunlight in the same way as carbonaceous asteroids, so it is suggested that all four are fragments of a carbonaceous asteroid that broke up on capture by Jupiter.

Three groups defined by common reflectance characteristics and similar orbital radii, eccentricities, and inclinations are recognized among Jupiter's retrograde irregular satellites. They are each named after their largest member: the seven-strong Ananke group has orbits near 297 Jupiter radii, the thirteen-strong Carme group near 327 Jupiter radii, and the seven-strong Pasiphae group near 330 Jupiter radii. These are believed to be fragments of asteroids of other types. Many of Jupiter's irregular satellites, including more than twenty beyond the Pasiphae group, have no known families. Each could be either a small captured asteroid or comet nucleus.

Quite when these moons were captured is unknown. It would have been much easier to achieve very soon after Jupiter had formed, because it could then have had an extended diffuse atmosphere to provide the necessary drag to slow the incoming objects down sufficiently to become captured. Nor is it clear when the parent objects for each related group broke up. It could have been during the capture process, or afterwards because of a collision.

None of Jupiter's irregular satellites shows synchronous rotation. They are too small and too far from Jupiter for tidal forces to be effective, and this is borne out by the few examples where rotation periods have been measured and shown to be only a few hours, in contrast to orbital periods of hundreds of days. On the other hand they feel the Sun's pull so strongly that the shapes and inclinations of their orbits can vary markedly in only a few years.

The irregular satellites of the other giant planets follow a similar pattern to Jupiter's, and their origins are probably similar. The most distant moons are all retrograde, but closer to the planet prograde and retrograde moons are intermingled rather than being neatly segregated like they are at Jupiter.

Saturn has thirty-nine known irregular moons counting Hyperion (Figure 11), a 180 × 133 × 103 km radius moon that has an orbit

11. Hyperion (above) and Phoebe (below) seen at similar scales.

between those of the outer two regular satellites, Titan and Iapetus. Hyperion is unique among known moons in that its rotation is chaotic. Not only is its rotation period variable, but even its axis of rotation changes as it tumbles along. Its density is only half that of solid ice, and it probably has a porous, rubbly interior. It has a low albedo, suggesting a surface dusting of dark particles which is typical of this part of the Saturnian system.

Phoebe (Figure 11) is 109 × 109 × 102 km in radius and is the largest and closest of Saturn's retrograde irregular moons. Phoebe's orbit lies at 548 Saturn radii. This is too far out to be visited by Cassini after it had achieved orbit, so Cassini's approach to Saturn was timed to enable it to make a close pass by Phoebe on its way in, at a range of 2,000 km, making it the best-imaged example of all its kind. Cassini revealed a cratered surface and detected water ice, carbon dioxide ice, and clay minerals. Phoebe has an extremely low albedo, only 0.06, which may be because methane ice has been stripped of some of its hydrogen (long exposure to the Sun's ultraviolet radiation can do this), allowing the carbon atoms to link together as a black tarry goo. Phoebe is a good candidate to be a captured centaur—a class of icy asteroid found mostly beyond the orbit of Saturn. In 2009 infrared telescope observations revealed that Phoebe orbits within a diffuse but very broad (twenty times thicker than Saturn itself) belt of dust, thought to have been knocked off Phoebe's surface by micrometeorite impacts.

Saturn has two orbital groups of prograde irregular satellites. The individuals in one are given Inuit names such as Siarnaq, and in the other Gallic names such as Albiorix. Each group could be the remains of a larger moon destroyed by collision. Apart from Phoebe, Saturn's retrograde irregular satellites have Norse names, and include groupings that could each represent fragments of the same captured asteroid.

Irregular moons of Uranus and Neptune are challenging objects even for the best modern telescopes. Uranus has nine, discovered in the period 1997–2003. They have retrograde orbits except for Margaret, which has the most eccentric orbit of any planet's moon. The largest is Sycorax, about 75 km in radius, whereas the smallest known have radii of about 10 km. There are no close orbital groupings, and each is probably an individually captured object.

Only six irregular moons of Neptune are known: three prograde and three retrograde. The largest, Nereid, has a radius of 170 km. It was discovered in 1949, and the others in 2002–3. The outermost examples, Psamathe (20 km radius) and Neso (30 km radius), are in eccentric retrograde orbits at mean distances of 1,885 and 1,954 Neptune radii. This is a vast distance (they take over 9,000 days to orbit the planet), but these orbits are stable because Neptune's Hill sphere is larger than Jupiter's, thanks to its greater distance from the Sun.

Nereid may be a large surviving remnant of a regular satellite that was catastrophically destroyed (maybe in the Triton capture event). It has been shown to have water ice on its surface, with an imposed low albedo thanks to a darkening agent such as carbon, in which respect it resembles some of the regular satellites of Uranus. It has been proposed that one other irregular satellite, Halimede, could be a smaller fragment from the same body, but Neptune's other irregular moons are probably individually captured objects.

Inner moonlets

Inner moonlets can be very small, as their name implies, and their proximity to the glare of their planet makes them harder to detect by telescope than irregular satellites. There is a good reason why large moons are not found close to their planets, articulated by the French astronomer Édouard Roche (1820–83) who calculated the distance from each planet at which the difference between the planet's tidal pull on the moon's near and farsides would exceed the moon's own gravity. At this distance, commonly referred to as the 'Roche limit', a fluid or loosely consolidated body would be pulled apart, though the internal strength of a solid body allows it to approach closer before it disintegrates.

Most inner moonlets orbit within their planet's Roche limit, and are probably fragments of larger moons ripped apart by tides. Some of the more distant, and larger, examples may have originated as regular satellites battered by collisions.

Only four inner moonlets of Jupiter are known, of which only the largest, Amalthea, had been discovered before Voyager. The best Galileo images of these are included in Figure 12.

Saturn has eight known inner moonlets within the orbit of Mimas, and there are three others (only about 1 km in radius) between Mimas and Enceladus. Of those three, only Methone has been seen at close quarters by Cassini, and has a surprisingly smooth egg shape. The best Cassini images of Saturn's conventional inner moonlets are included in Figure 12, and most show much better detail than provided by Galileo at Jupiter. Saturn's closest inner moonlets, Pan and Atlas, have ridges running round their equators, giving them curious 'flying saucer' shapes. Their orbits follow gaps in Saturn's ring system, and the equatorial ridges are probably swept up ring material.

Janus and Epimetheus are a unique pair: the orbit of one is only a few km bigger than the other. When the inner, faster travelling, moon catches up with the other, which happens about every four years, their mutual gravitational interaction causes them to swap orbits. The previously faster one is now in the wider orbit so it slows down, whereas the previously slower one speeds up until it catches up with its fellow four years later and the cycle repeats.

Uranus has thirteen known inner moons, of which only Puck (Figure 12) was seen in any detail by Voyager 2, which discovered

12. A compilation of the best-imaged inner moonlets, shown at approximately correct relative scales (Pandora is 100 km from end to end). For Jupiter (Galileo images) and Saturn (Cassini images), moonlets are arranged with closest to the left. Two views of Atlas are included: equatorial (above) and southern hemisphere (below). Puck and Proteus images are from Voyager 2.

it. They are all dark (low albedo) objects, probably as a result of radiation damage to methane. Most were discovered by Voyager 2, but three (including the smallest, Cupid and Mab, which are only about 5 km in radius) are more recent telescopic discoveries. All thirteen are crowded into a narrow orbital range, 1.95 to 3.82

Uranus radii, and simulations suggest that they can disturb each other's orbits so that there is likely to be a collision sometime in the next 100 million years.

Six inner moonlets of Neptune were revealed on Voyager 2 images, and a seventh (the smallest, at about 18 km radius) was discovered by imaging with the Hubble Space Telescope. Like their equivalents at Uranus, they have low albedo, and seem to be water ice made dark by radiation-damaged methane. One of these, Proteus (Figure 12), is the largest inner moonlet in the Solar System, at 220 × 208 × 202 km radius. Its shape is a long way from hydrostatic equilibrium and probably results from collisional battering. The innermost of Neptune's moonlets, Naiad, orbits well inside the Roche limit, and its eventual fate will probably be either to spiral into the planet's atmosphere or to be ripped apart by tides and make a new ring.

Rings and shepherd moons

Saturn is famous for its spectacular ring system. This was first seen by Galileo, but he couldn't make it out clearly. Huygens (in 1655) was the first to correctly interpret what he saw. The other giant planets have rings too, though these contain much less mass and so are far less spectacular.

Even Saturn's glorious rings contain less mass than Mimas, its smallest regular satellite. They are probably very old, and represent mostly either material that was too close to the young planet to have ever been gathered into a moon, or the remains of a moon that was ripped apart when it strayed within the Roche limit. The easily visible extent of the rings (which you can see for yourself with a small telescope) consists, working outwards, of rings designated C, B, and A by the IAU. These are only a hundred metres thick and extend from about 75,000 km to about 137,000 km from the planet's centre. That's 1.24 to 2.27 Saturn radii, but of course only 0.24 to 1.27 Saturn radii above the cloud tops. Spectroscopy shows that these rings are mostly water ice, darkened and in most places reddened by radiation damage or dusty contaminants. The relatively slow cooling rate of the rings when they pass into Saturn's shadow, the ability of ground-based radar to record a signal reflected from the rings, and the effect of the rings on Voyager 1's radio transmissions show that the rings are made of chunks ranging from about one centimetre to five metres in size. Each such chunk is in orbit about the planet. It would be perverse to regard every one of them as a moon, though there is no agreed lower size limit for what can be called a moon.

Less substantial rings occur to either side of the main rings. The D ring (66,900–74,500 km) lies between the C ring and the planet, whereas other rings lie beyond the A ring. Saturn's rings have amazingly complex internal structures, most obviously at distances from Saturn where the ring material becomes more diffuse or even vanishes completely. The widest 'gap' is 4,800 km wide and separates the A and B rings. It was discovered in 1675 by Giovanni Cassini and known as the Cassini Division. Its inner edge occurs at the distance where any ring particles would have twice the orbital period of Mimas, and so is readily explicable. Fine structure within the gap, at radii where particles are either concentrated or dispersed, is not fully understood.

Numerous minor gaps elsewhere in the rings defy simple orbital resonance explanation, but some are swept clear by inner moonlets whose orbits actually follow the gaps—for example, Pan and Daphnis each sweep a different gap in the A ring (Figure 13).

Cassini images in 2013 showed a 1,200 km long, 10 km wide enhancement of material at the outer edge of the A ring, which might be concentrated around a small mass, perhaps 1 km across. This could be a new moonlet in the making, but is more likely a temporary disturbance that will disperse.

13. Moons among the rings. Top: Prometheus, 176 km long, travelling from right to left, making periodic disturbances in Saturn's F ring. Middle left: Pan, 34 km long, orbiting in the Encke gap, casting long shadow thanks to grazing-incidence sunlight. Middle right: Daphnis, 8 km long, orbiting in the Keeler gap, making waves in the inner edge of the A ring. Bottom: Uranus' rings with two moons circled, Cordelia (left) and Ophelia (right). Each is about 20 km in radius, and they appear artificially elongated in this long-exposure image. There is a lot of speckly noise, and the dark band either side of the outermost ring is an artefact.

The F ring is only 30–500 km wide and lies outside the A ring. This is held in place, and given a dynamic structure, by the inner moonlets Prometheus (Figure 13) and Pandora that orbit just inside and just outside it, respectively. Inner moonlets such as these, whose intimate association with a ring system helps to define its shape, are often referred to as 'shepherd moons'.

Beyond the F ring there are various tenuous rings and ring arcs believed to be dust knocked off the outermost inner moonlets, Aegaeon, Methone, Anthe, and Pallene, by micrometeorite bombardment, and confined or concentrated into arcs by resonance with Mimas. Further out still, at 180,000 to 480,000 km, is the diffuse E ring consisting of particles less than a micrometre across made of water ice and traces of other material erupted from Enceladus. The belt of diffuse dust that envelopes Phoebe's orbit is sometimes referred to as an additional ring, but is far too thick for the term to be appropriate. A suggestion dating from 2008 that Saturn's regular satellite Rhea is encircled by its own diffuse ring system has now been discounted. In fact, no moon is known to have rings, just as no moon has a moon.

Jupiter's rings were discovered by Voyager 1. The main ring is only 6,500 km wide and 30–300 km thick. The orbit of the innermost inner moonlet, Metis, occupies a gap within this ring, and the ring's outer edge is shepherded by the orbit of Adrastea. The ring material is an unknown reddish material, consisting of a mixture of micrometre-sized dust and larger chunks, with a total mass no more than about five times that of Adrastea and possibly much less. It would take less than 1,000 years for radiation pressure and interactions with Jupiter's magnetic field to disperse the dust, so unless it is extraordinarily young it must be being replenished either by collisions between the larger chunks, or by micrometeorite bombardment of Metis and Adrastea.

NASA's New Horizons mission searched in vain for unknown inner moonlets within this ring during its 2007 fly-by, so there are unlikely to be any greater than 0.3 km in radius. It did however find seven clumps of material, occupying arcs extending for 1,000 to 3,000 km.

Inward of the main ring, Jupiter has a 12,500 km thick 'halo ring', which seems to be dust spiralling inwards to Jupiter from the main ring. There are also two exceedingly faint 'gossamer' rings, which are dusty features extending outwards from the orbits of Amalthea and Thebe, the outermost inner moonlets, which are probably supplied by ejecta from micrometeorite bombardment of their surfaces.

The rings of Uranus were discovered in 1977, by telescopic observers who noted the repeated dimming of a star as the otherwise invisible rings passed across it. Most of our knowledge of the rings comes from Voyager 2 and the Hubble Space Telescope. Thirteen rings are now known, with a total mass exceeding Jupiter's rings but much less than Saturn's. Several are narrow and consist mostly of boulder-sized chunks of low-albedo reddish material, believed to be produced by the fragmentation of inner moonlets. These are all narrow rings, the outer and brightest of which (the epsilon ring) is shepherded by the moonlets Cordelia and Ophelia (Figure 13). It is at a mean distance of 51,150 km from the planet's centre, but is eccentric in shape so that the distance varies by 800 km around its circumference. It is 20 km wide where closest to the planet and 100 km wide where furthest. The inner rings are even more tightly confined but have no known shepherd moons, prompting the suggestion that these formed by fragmentation no more than about 600 million years ago.

Uranus' other rings are dusty. There are four inside the narrow rings, believed to be short-lived dust spiralling towards the planet, and two beyond—the outermost of which is spread round the orbit of the outermost inner moonlet Mab, which is probably the source of its dust.

Neptune has five rings, which are extremely dark, like those of Uranus. They are about half dust and half larger chunks. Material in the outermost ring is concentrated into several arcs. Attempts to explain this as a 42:43 resonance with the inner moonlet and ring shepherd Galatea have failed, and it is clear that there is much about rings and their relationship with moons that we do not yet understand.

Chapter 5
Regular satellites in close up

Twenty-five years ago I wrote a Voyager-based book describing the regular satellites of the giant planets as 'worlds in their own right'. The truth of that phrase is even more evident now. Each of them is individually fascinating, and two are even widely regarded as better candidates than Mars for hosting extraterrestrial life. Here I will discuss some of my favourites.

Io

Thanks to tidal heating, Io is the most volcanically active body in the Solar System, surpassing even the Earth. Nine eruption plumes were discovered by the Voyager probes, of which the largest was 300 km high and more than 1,000 km wide. The Voyagers detected strong infrared radiation from the plume sources demonstrating high temperatures, and increasingly sophisticated infrared telescopes on Earth documented these and many other 'hot spots' on Io during the sixteen-year interlude between Voyager and Galileo, and in the periods before and after the 2007 fly-by by New Horizons.

The discovery of active volcanoes on Io led the IAU to revise the naming convention that had been lined up for features on Io, and to choose instead names of mythical blacksmiths and fire, sun, thunder, and volcano gods and heroes. For example, of the erupting volcanoes in Figure 14, Prometheus is a Greek god who stole fire to give to humans, Tvashtar is named after a sun god in Sanskrit epics, and Masubi is a Japanese volcano deity. Eruptions have now been documented at about a hundred sites, of which a few (such as Prometheus) are persistent and some others (such as Tvashtar and Masubi) have erupted on several occasions.

Io lacks any visible impact craters, because it is continually being resurfaced by volcanism that rapidly buries craters. This resurfacing occurs by a combination of fallout from eruption plumes and over-riding/flooding by lava flows. The strong yellow and orange colours that dominate Io's surface were originally thought to show that many or all of the lava flows were made of sulfur, but infrared measurements of the active vents show that the temperature within them is far too hot for sulfur. The hot material must be molten rock, probably a low-viscosity magma low in silica and rich in magnesium.

Low-albedo lava flows are recognizable elongated lobate features up to 300 km long, and often emerge from caldera-like landforms whose circumferences are neither smooth nor circular enough to be impact craters. After a lava flow has ceased to move, it begins to fade from visibility as it becomes coated with sulfur and sulfur dioxide frosts (responsible for Io's colouring), which are both exhaled from the ground and deposited as fallout from eruption plumes.

Some calderas have no flows emerging from them but are strong sources of infrared radiation, and evidently contain 'lava lakes' where magma continually wells up, crusts over, and sinks to be replaced by fresh magma from below.

Other volcanic vents or fissures are the sources of Io's largest eruption plumes. These are presumably powered by the expansion of gas bubbles in magma as it rises through the crust, in which the gas expands with sufficient violence to shatter the magma to

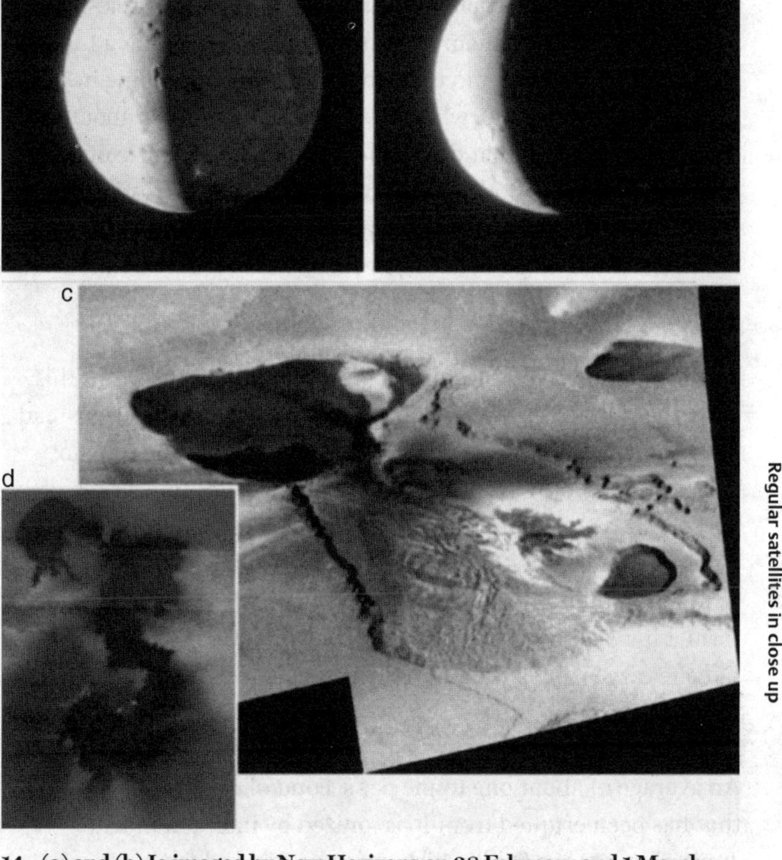

14. (a) and (b) Io imaged by New Horizons on 28 February and 1 March 2007. The 290 km high eruption plume from Tvashtar is prominent in the north. The 28 February image also shows a 60 km plume from Prometheus on the limb in the nine o'clock position, and the top of a third plume, from Masubi, rises high enough to catch the sunlight in the south. The nightside of Io is lit by light reflected from Jupiter, but the dayside has been overexposed to increase the plumes' visibility. On the 1 March image, the vent at Tvashtar can be seen glowing in the dark. The Masubi plume can be seen above the dark limb in the four o'clock position; c 300 km wide mosaic of Galileo images of Tvashtar as it was on 16 October 2001. Shadows change confusingly because images from different times of day have been combined; d 90 km wide mosaic of Galileo images of lava flows erupted from a caldera (at top left). The Prometheus plume originates near the front of the main lava flow, which is towards the bottom of the image.

fragments and propel it skywards. On Earth, this process would draw in the surrounding air and develop into a convecting column, but Io has virtually no atmosphere so the particles follow parabolic arcs taking them skywards and then back down to the ground. This is well demonstrated by the Tvashtar plume in Figures 14(a) and 14(b), where the eruption speed was about 1 km per second. The gas that drives the eruption is a mixture of sulfur and sulfur dioxide. Both substances condense into 'frost' particles near the top of their trajectories, and fall to the ground along with the fragmented magma, but in this case the magma dominates so that the deposit has a low albedo.

The Prometheus and Masubi plumes are a different sort. In this case, the plume does not originate at a volcanic vent but instead at the advancing front of a lava flow. In this situation, the hot lava vaporizes the sulfur dioxide surface-frost of the over-ridden surface. It is this that forms the plume, which becomes visible as the expanding gas condenses to frost particles. Plumes of this kind are mainly sulfur dioxide with little or no magma, and form a high-albedo deposit where they fall. Many such white blotches are visible in Figure 14(a), especially in the Jupiter-lit nightside area.

An average of about one tonne per second of oxygen and sulfur that has been erupted from Io is ionized by interaction with Jupiter's magnetic field (which envelops Io), and feeds a doughnut-shaped 'plasma torus' that encircles the planet at the radius of Io's orbit.

Europa

Europa could scarcely look more different to Io (Figure 15), though if we could strip away about 100 km of ice and water we might reveal a less active version of Io beneath. Europa's surface is high-albedo ice, too young to have accumulated more than a few impact craters. Much of the surface is a mass of ridges and grooves,

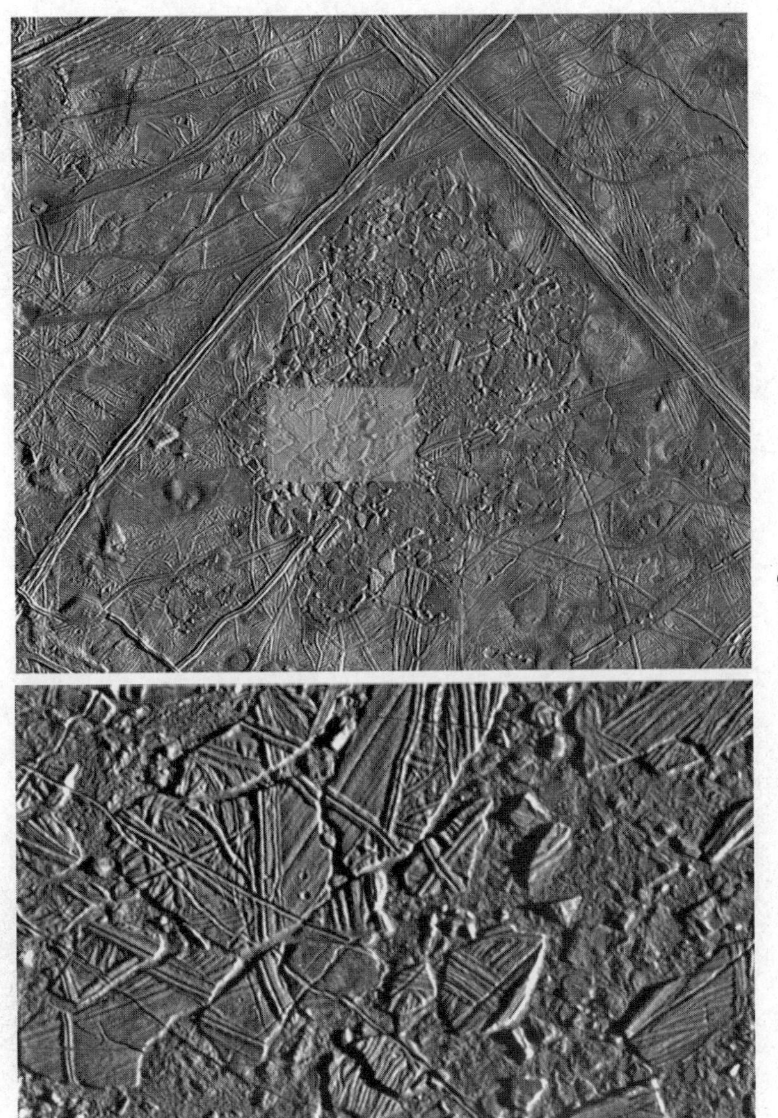

15. Europa imaged by Galileo. The upper image is a 300 km wide area, showing ball of string terrain, except centrally where it has been broken apart to form chaos terrain. The area in the transparent rectangle is shown below at higher resolution, revealing rafts and matrix. The two largest impact craters in the lower image are less than 200 metres in diameter, just left of centre.

contributing to an appearance that has been likened to a ball of string. Most ridges seem to be where a crack has opened and then closed again, squeezing out some slush to make a ridge at the surface. This opening and closing could be a tidal process, and so would repeat with each eighty-five-hour-long orbit round Jupiter, building a hundred-metre-high ridge of refrozen slush over a crack's active lifetime before it seals for good and a new crack opens elsewhere. Some particularly large ridges are fringed by discoloured ice, demonstrating impurities (probably magnesium sulfate and other salts) in the water from which the slush froze.

The ball of string terrain is disrupted in various regions, mostly by being broken apart to form a landscape described as 'chaos'. Here, you can see rafts with ball of string textures separated from each other by a lower-lying surface that has a fine jumbled texture. Most scientists now accept that this 'matrix' between the rafts is the refrozen surface of a body of water that had been temporarily exposed when the surface ice became thin, allowing it to break into rafts that drifted apart. In some chaos regions it is possible to see how the rafts once fitted together, but in others whole rafts have vanished—either sunk or melted. In older chaos regions, some new ridges and grooves cross rafts and matrix alike, suggesting that eventually chaos becomes unrecognizably overprinted into a new generation of ball of string terrain.

Even very salty water would freeze quickly upon exposure to space at Europa's –160°C surface temperature, but we can imagine kilometre-thick ice rafts drifting through an ocean surfaced by tens of metres of slush for many weeks before the slush became too thick and stiff to permit further drifting. Calculations based on the typical hundred-metre height difference between raft surface and matrix surface, and allowing for the likely buoyancy difference between raft and salty water, suggest that when the ice sheet broke into rafts it was between half a kilometre and a few kilometres thick.

One thing that experts don't yet agree on is whether the body of water exposed when each chaos forms is the top of Europa's subsurface global ocean (Figure 9), or merely the top of an isolated lens of liquid water within Europa's ice shell. We have no way of telling the thickness of this ice shell (a Europa orbiter to map the height of the tidal bulge and equipped with ice-penetrating radar would be the best way). The shell could be tens of kilometres thick, in which case the melt-through process to thin the shell upwards from its base until it was thin enough to break apart would require a sustained localized input of heat capable of melting tens of kilometres of ice, whereas if the shell is only a few kilometres thick melt-through would be much easier to achieve. The heat to drive this process would be tidal heat generated in Europa's rock interior, and possibly delivered to the floor of the ocean by a submarine volcanic eruption.

Proponents of the thick-ice model of Europa prefer a temporary lens of liquid water to be melted within the ice, by heat transported upwards by solid-state convection within the deeper ice. This could either be powered by tidal heating within the ice, or be a response to a submarine volcanic eruption conveying heat to the ocean.

Even though it has essentially no atmosphere, Europa is a more promising, though less convenient, place to search for extraterrestrial life even than Mars. Life on Earth is believed to have begun at 'hydrothermal vents' on the ocean floor, and the tidal heating in Europa's rocky shell that supplies the heat to keep the subsurface ocean from freezing should be associated with hydrothermal vents aplenty, where water that has been drawn inwards through the rock is expelled after heating.

There are ecosystems in Earth's deep oceans that survive on the chemical energy emerging from hydrothermal vents. These do not depend at all on sunlight, which plants use to power their metabolism. It is entirely feasible that something similar exists round vents on Europa's ocean floor, where they might be 'chemosynthetic' microbes at the base of the food chain, and larger, possibly multicellular, predators feeding on them.

If life began at deep-sea vents on Europa, it may have subsequently adapted to colonize other environments that would be easier for us to explore. The best would be a tidal crack—up which water would have been drawn each time it had opened. Any planktonic organisms that reached the upper few metres would find enough sunlight to allow photosynthesis, and provided that they stayed at least a few centimetres below the surface the water would shield them from the severe radiation dose that would be experienced at the surface. Some unlucky organisms would be squeezed out with the slush each time a crack closed, so the easiest place to find fossilized life would be in entombed in the frozen icy slush that makes up a surface ridge.

Ganymede and Callisto

Ganymede may also have an ocean resting on rock, which is an important factor for the supply of chemical energy for any life to feed on. As suggested in Figure 9, this may be overlain by several successive solid and liquid shells. However, whatever the exact situation, the top of even the topmost ocean is deep down and remote from the surface. Ganymede has no recognizable traces of chaos terrain, so maybe its surface ice shell never became thin enough for melt-through to occur. However, there have been multiple episodes in which parallel swarms of grooves were formed. The youngest examples, which cross-cut older terrain, have higher albedos than other areas (Figure 16). This might be Ganymede's local equivalent of ball of string terrain, but most grooves are cracks without the ridges seen on Europa.

An important contrast with Europa is that Ganymede's surface fracturing seems to have ceased long ago. Figure 16 shows numerous craters from 2 km downwards on the belt of high-albedo, youngest

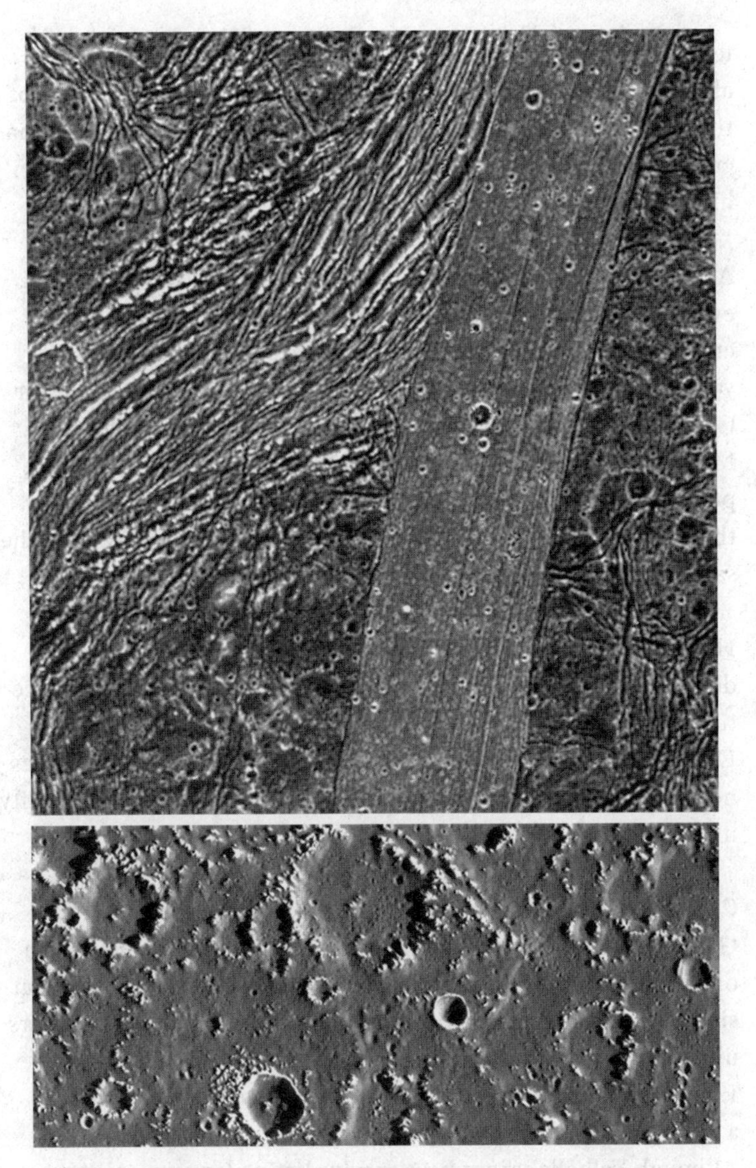

16. Representative 70 km wide regions of Ganymede (top) and Callisto (bottom), imaged by Galileo. The youngest terrain on the Ganymede image is the high-albedo belt of fine grooves running north–south, but even this must be very old because there has been time for many impact craters to be made on it. Note that the Sun was high in the west for the Ganymede image, but lower and in the east for the Callisto image.

terrain. The older, darker terrain has craters like this too, but it also has craters up to 10 km in size that have been cut by many of the cracks. Beyond the region shown in Figure 16, Ganymede has craters up to 200 km in size, some of which are superimposed on high-albedo belts.

Although the crater density on Ganymede's surface is convincing evidence that the surface is old, sadly we cannot apply the lunar cratering timescale to derive an absolute age. This is because if you plot the distribution of crater frequency versus crater size for Ganymede, it looks very different to the equivalent plot for the Moon. This shows that Ganymede has been struck by a different population of impactors, and that there is no reason to assume that the history and rate of bombardment at Jupiter have been the same as at the Earth, five times closer to the Sun.

However, we *can* conclude that the dramatic contrast in crater densities between Ganymede and Europa provides clear evidence that the average age of Ganymede's surface is much older than Europa's. It could easily be older than three billion years, whereas no surviving surface on Europa is likely to be older than about fifty million years.

Callisto has an even more heavily cratered surface than Ganymede, and if it ever did have surface cracks these have been obliterated by the cratering. Figure 16 shows an area of Callisto at similar scale to the accompanying view of Ganymede, with craters up to 15 km wide, but there are plenty of bigger craters elsewhere. In fact Callisto has the largest impact crater in the Solar System, a 3,800 km diameter multiple-ringed basin named Valhalla. Although Valhalla's rings have survived, it no longer retains its original depth and has become flattened because Callisto's ice was unable to sustain such a deep basin, and flowed upwards to repair the damage. Callisto's internal ocean (Figure 9) is inferred from its induced magnetic field. It has had no discernible effect of the surface and must lie much deeper than Europa's ocean.

Titan

Although Saturn has seven regular satellites, Titan is the only one to rival Jupiter's Galilean moons for size. It is only slightly smaller and less massive than Ganymede (see Appendix), and has an icy outer shell, which Cassini observations showed to be separated from the interior by an internal ocean beginning at a depth of about 100 km. If this ocean is water mixed with ammonia it could be about 200 km thick, and below it would be a 200 km thick layer of Ice VI surrounding Titan's rocky interior.

The thing that makes Titan particularly special is that, unique among moons, it has a dense atmosphere, whose surface pressure is about one and a half times that on Earth. It is about 98 per cent nitrogen with most of the rest being methane (CH_4). Between 50 and 250 km above ground level, solar ultraviolet light splits hydrogen atoms off from methane (a processes called photodissociation), whereupon the carbons link into chains to make a hydrocarbon haze that hides the surface from view at most wavelengths of light. The Voyagers saw nothing of the surface, but Cassini carried an imaging radar specifically to penetrate this haze and map the terrain. In addition, Cassini's optical camera was able to obtain low-resolution views of the surface using a special filter, which has proved useful for tracking seasonal surface changes. We have also seen the surface from the Huygens probe (carried to Titan by Cassini), which sent back clear pictures from below the haze during its parachute descent and from the ground itself.

And what a surface was revealed! If you did not know that the surface 'rock' is actually water ice, and the few fluffy low altitude clouds are condensed methane and/or ethane, you might think you were looking at a part of the Earth. Titan experiences methane rainfall, which collects in streams and rivers to erode valleys, which drain into lakes, three of which are large enough to be called seas, such as Kraken Mare in Figure 17. These seas and most of the lakes are at high northern latitudes, but there are lakes

17. Map-projected 450 km wide area near Titan's north pole, compiled from Cassini imaging radar swaths of variable quality collected in the years 2004–13. North is to the left. The 80° N line of latitude is superimposed, plus lines of longitude at 10° intervals. The large dark area is Ligeia Mare, less than five metres deep near its northern shore allowing radar to reflect from the seabed, but in excess of 100 metres deep elsewhere. Several rivers can be seen draining into it.

near the south pole too. Dry lakebeds have also been identified, with some signs of seasonal changes. Evidence of changing lake volumes comes also from the characteristics of their shorelines; for example, much of the southern shore of Ligeia Mare looks like a typical drowned coastline where hills and valleys carved on dry land have been flooded by a rise in sea level. There is also the possibility of spring-fed ponds and lakes. The liquid seeping into them would have begun as methane rain that seeped into the icy crust and percolated underground (through what on Earth would be called an 'aquifer'), where some of it could be converted to ethane or propane before emerging again.

The Huygens lander touched down only a little way south of the equator. It saw dry river valleys on its way down, and landed on a flood plain or dry lakebed. The area was strewn with pebbles of ice that had been rounded by transport along a methane river, sitting on a dark substrate that could be a tarry mix of rained-out hydrocarbons.

Vast fields of low-albedo sand dunes cover the icy 'bedrock' at low latitudes, with individual dunes about a hundred metres high and tens of hundreds of kilometres long. These dunes appear to be formed parallel to eastward-blowing winds, driven by tidal forces from Saturn. The 'sand' grains in these dunes could be dirty grains of ice, or hydrocarbon particles.

Despite the depth and density of Titan's atmosphere, large impactors should be able to reach its surface at the hypervelocities needed to make impact craters, but only eight impact craters have been identified, ranging from 29 to 292 km in diameter. This paucity of craters shows that the average surface age is young, but we do not know whether the older craters have merely been erased by erosion (rivers and wind) and buried by sediment, or whether other processes are at work too. One of these could be volcanism. There are a few candidates for volcanoes on Titan, such as Sotra Patera, a twin-peaked mountain 1.5 km high, with craters from which apparent lava flows have emerged. The lavas in this case would not have been molten rock, but something derived by the partial melting of Titan's icy crust. It could have been an ammonia-water mix (which is liquid at a much lower temperature than pure water), or even a waxy material derived by processing buried hydrocarbons that originally rained out of the atmosphere.

To distinguish it from conventional volcanism involving molten rock, as on Earth, the Moon, and Io, volcanism in which the erupted material is derived from ice of any kind is called cryovolcanism. On Titan, significant tidal heating seems unlikely, so the driver may be heat from radioactive decay in the rocky interior.

Enceladus

Enceladus is the smallest of Saturn's regular satellites apart from its inner neighbour Mimas. However, whereas Mimas is covered in impact craters, Enceladus is a much more interesting place. Voyager showed cratered terrain juxtaposed against terrain that looked smooth and featureless. The higher-resolution imaging achieved by Cassini revealed that the apparently smooth areas are in fact intensely fractured, and that the impact craters that used to be there have been largely erased by multiple generations of cross-cutting fractures.

Even more exciting was Cassini's discovery of plumes being erupted from a series of fissures near the south pole (Figure 18), that became colloquially known as the tiger stripes because the surface along them shows up as bands of slightly bluer, fresher ice. More than a hundred individual geysers have been identified, which jet tiny crystals of water ice into space as speeds of about 1 km/s. These are evidently the source of the material forming Saturn's E ring. The plumes were not anticipated when Cassini was designed, but the mission plan was revised to allow the craft to swoop through them at ranges of 200 to 25 km above the surface. Cassini's Ion and Neutral Mass Spectrometer (which had been intended primarily as a tool for sampling Titan's exosphere

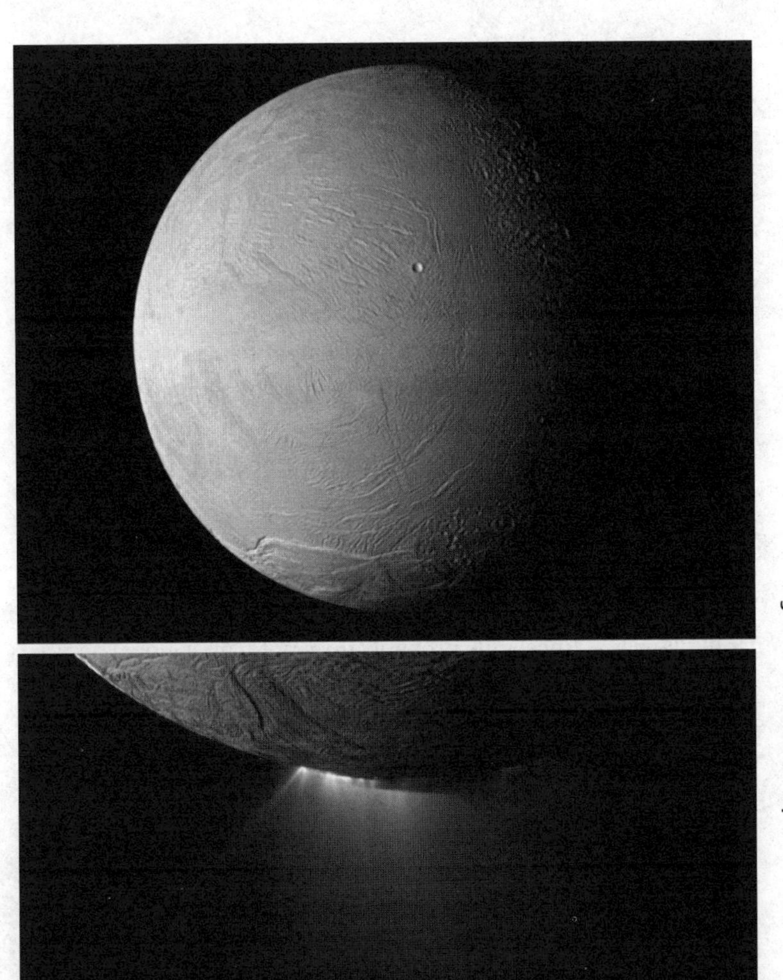

18. Enceladus seen by Cassini on 21 November 2009 (top); south polar region imaged by Cassini on 30 November 2010 (bottom), showing sunlit plumes emerging from three sub-parallel fissures.

and ions, atoms, and molecules associated with Saturn's magnetosphere) showed that the plume is 99 per cent water with traces of methane, ammonia, carbon monoxide, carbon dioxide, and various simple organic molecules. Cassini's infrared spectrometer was able to show that the temperature inside the ten-metre-wide fissures from which plumes erupt is at least as warm as –70°C, in contrast to the regional surface temperature of –180°C. Enceladus' orbit is only very slightly elliptical, and the plumes are strongest when Enceladus is at its furthest from Saturn, which is when the stresses in the crust would tend to pull the tiger stripe fissures open.

The source of the vented material is thought to be a vast pod of melted water below the south pole, warmed by tidal heating. Tidal cracking is presumably responsible for the vastly complex history of surface fracturing exhibited across the globe. It is far from certain that the present-day rate of tidal heating is sufficient to sustain the current activity, which may be left over from a previous episode of stronger heating.

Unlike Europa, Enceladus probably does not have a global ocean and nor is the liquid water pod likely to be resting on rock. The small extent and the limited chemical interaction between water and rock means that the potential for Enceladus to host microbial life is less than for Europa. However, the delivery of samples to space where they can be accessed without landing means that a future fly-by or orbiter probe with specially designed instruments could search for organic molecules or biologically driven fractionation of isotopes, without needing to land on the surface.

Before Enceladus' plumes had been discovered, the Galileo mission searched for eruption plumes from Europa from 1995 to 2003 without success, and nor were any seen during Cassini's 2001 fly-by of Jupiter. In 2012 the Hubble Space Telescope detected a diffuse zone of atomic oxygen to one side of Europa that is most simply explained as a product of photodissociation of water vented from a plume. However, repeat attempts at detection throughout 2014 were blank, and if eruption plumes do indeed occur at Europa it seems clear that they must be weaker and less persistent than those currently occurring at Enceladus.

Irrespective of whether or not there is life in the liquid water zones of Europa and Enceladus, it is conceivable that the right kind of terrestrial microbes could survive and reproduce if they found themselves there. Space probes are cleaned and sterilized before launch, but it is impossible to remove or kill all the microbes, and some can survive for years in space. If probes such as Galileo and Cassini were left as derelict hulks in orbit, there is a chance that one day they might crash into Europa or Enceladus, unintentionally delivering viable microbes from Earth. To prevent this, at the end of its life in 2003 Galileo was deliberately crashed into Jupiter, and Cassini will meet a similar fate at Saturn in 2017.

This is not just a matter of the ethics of protecting alien ecosystems, though arguably that should be an important consideration behind any so-called 'planetary protection' protocols. We also want to study any such ecosystems free of the confusion that any spacecraft stowaways would engender. If these ecosystems exist, it will be important to establish whether life began there independently of life on Earth, or whether life within the Solar System shares a common origin, perhaps having been spread from world to world as accidental passengers inside meteorites.

There can hardly be any bigger question than that of whether life started independently more than once in our Solar System. Until we find unrelated life somewhere, life on Earth could be just an extraordinarily rare statistical fluke. But if life started inside an icy moon independently of life on Earth, then surely it has also begun on other habitable worlds throughout our galaxy. A dedicated mission to seek life at Enceladus or Europa might be the best chance we have of revising the answer to the question 'Are we alone?' from 'maybe' to 'no!'

Iapetus

Iapetus orbits Saturn at almost three times the distance of Titan, making it the most distant of any giant planet's regular satellites. It is also unusual in that its orbit is inclined at 15.5° to Saturn's equator, as a result of which it is the only regular moon of Saturn from whose surface the planet's rings would sometimes be clearly visible.

When Giovanni Cassini discovered Iapetus in 1671 he was able to see it only when it lay to the west of Saturn in the sky. It took him more than thirty years of trying before he glimpsed it faintly when it lay east of Saturn. Cassini correctly interpreted this as proof that Iapetus is in synchronous rotation and that its forward-facing hemisphere has a much lower albedo than the trailing hemisphere.

Thanks to the Cassini spacecraft, we now know that the surface distribution of albedo resembles the pattern on a tennis ball (Figure 19). A high (0.5–0.6) albedo tract runs from pole to pole via the centre of the trailing hemisphere. Its low (0.03–0.05) albedo counterpart is a broad equatorial tract centred on the leading hemisphere, and has been named Cassini Regio in honour of Giovanni Cassini. All other features on Iapetus are named after characters and places in the medieval French epic poem *Chanson de Roland*, which tells of a conflict between Franks and Saracens. With the exception of Cassini Regio, names in the high-albedo tract are generally Frankish and names in the low-albedo tract are generally Saracen.

Iapetus is very much a two-toned world, lacking shades of grey even at the most detailed image resolution of about thirty metres. This dichotomy could have been kick-started by Iapetus' leading hemisphere sweeping up dust sourced from impacts on Phoebe or smaller irregular satellites, but the sharp edges of the main dark tract and of small patches in the transition zone between the two tracts show that this cannot be the whole story. The dark material is probably a carbon-rich 'lag deposit' less than half a metre thick that has accumulated from impurities within the ice that were left behind while surface ice has very slowly sublimed away into space (which means passed directly from solid to vapour without

19. Iapetus seen by Cassini. The leading hemisphere (top); the trailing hemisphere (bottom). The albedo contrast between bright and dark sides is actually more extreme than it appears here.

melting). Iapetus' slow (seventy-nine days) rotation means that its surface has longer to warm up while facing the Sun than is the case for any other, faster spinning, regular satellite of Saturn.

Darker surfaces absorb more sunlight and heat up more than reflective surfaces, so that once an albedo contrast is established it will get stronger over time, until the non-volatile dark lag deposit is so thick that ice can no longer sublime from below it. Currently, the equatorial noontime temperature reaches about −144°C in Cassini Regio, whereas it is about 16°C colder in the high-albedo terrain.

Iapetus has an ancient, heavily cratered surface. Dark material sits on 'warm' crater floors whereas cold, pole-facing, inner walls of craters remain bright in the transition zone between high- and low-albedo tracts. Other peculiarities of Iapetus reside in its shape. Its polar radius is 34 km less than its equatorial radius. Allowing for Iapetus' low density, this degree of polar flattening corresponds to the equilibrium shape of a body rotating in only ten hours, and so the shape seems to have been frozen at a time before tidal drag had forced the rotation to synchronize with the orbit.

In addition, Iapetus has a narrow equatorial ridge about 13 km high, with peaks of up to 20 km. It can be traced for 1,300 km throughout Cassini Regio, where it can be made out in Figure 19, but it is present only as isolated 10 km peaks in the high-albedo tract. It must be ancient, because it is overprinted by numerous impact craters, and its origin is a mystery. It may be a feature inherited from Iapetus' formerly rapid spin. Alternatively, it may have something in common with the equatorial ridges on Atlas and Pan, but in this case it could not have been gathered from Saturn's rings and would have to be material accreted on to Iapetus' surface from a long-vanished ring of its own.

Miranda

Beyond Saturn, we come to giant planet moons visited only by Voyager 2 during fly-bys. Mission planners had a particular problem in the Uranus fly-by. Uranus' axis of rotation, and with it the orbits of its regular satellites, is tilted at 98° to its orbit.

Technically this means that Uranus' rotation is retrograde. The moons orbit in the same direction that the planet spins, so they count as prograde. However, the important point is that in January 1986 when Voyager 2 flew past, Uranus' south pole was facing more or less towards the direction from which the craft arrived, so the moons' orbits presented themselves like the rings around a target. It was not possible, as had been the case at Jupiter and Saturn, to cross each moon's orbit in turn and to time the encounters so that as many moons as possible lay in parts of their orbits close to where the probe passed. Instead, all that could be done to maximize the science at Uranus was to aim for a close fly-by of the innermost moon currently known, and to make the best of distant images of the other moons. Science was also limited because the northern halves of each body were in long-term seasonal darkness, and could not be imaged at all.

The innermost known moon was Miranda. This only has a 234 km radius and lacks present-day orbital resonance, so it was expected to be a fairly dull, heavily cratered object. However, it turned out to be startlingly complex and an absolute joy to behold (Figure 20). About half of the sunlit hemisphere has a uniform albedo and is densely cratered, but the older craters have muted profiles as if a layer of drab material has been draped over them, whereas the younger craters have fresh, crisp profiles.

The remainder of the sunlit hemisphere is taken up by three separate regions of lineated texture and (in two out of three cases) varied albedo. From left to right in Figure 20, these are Arden Corona, Inverness Corona, and Elsinore Corona, which take their names from places in Shakespearean plays. The IAU-approved descriptor term 'corona' signifies an 'ovoid-shaped feature', but does not help us understand what they are. The coronae have fewer impact craters than elsewhere, and they are all fresh looking, so clearly the coronae have younger surfaces than the rest of the globe. There is quite a lot of fracturing associated with

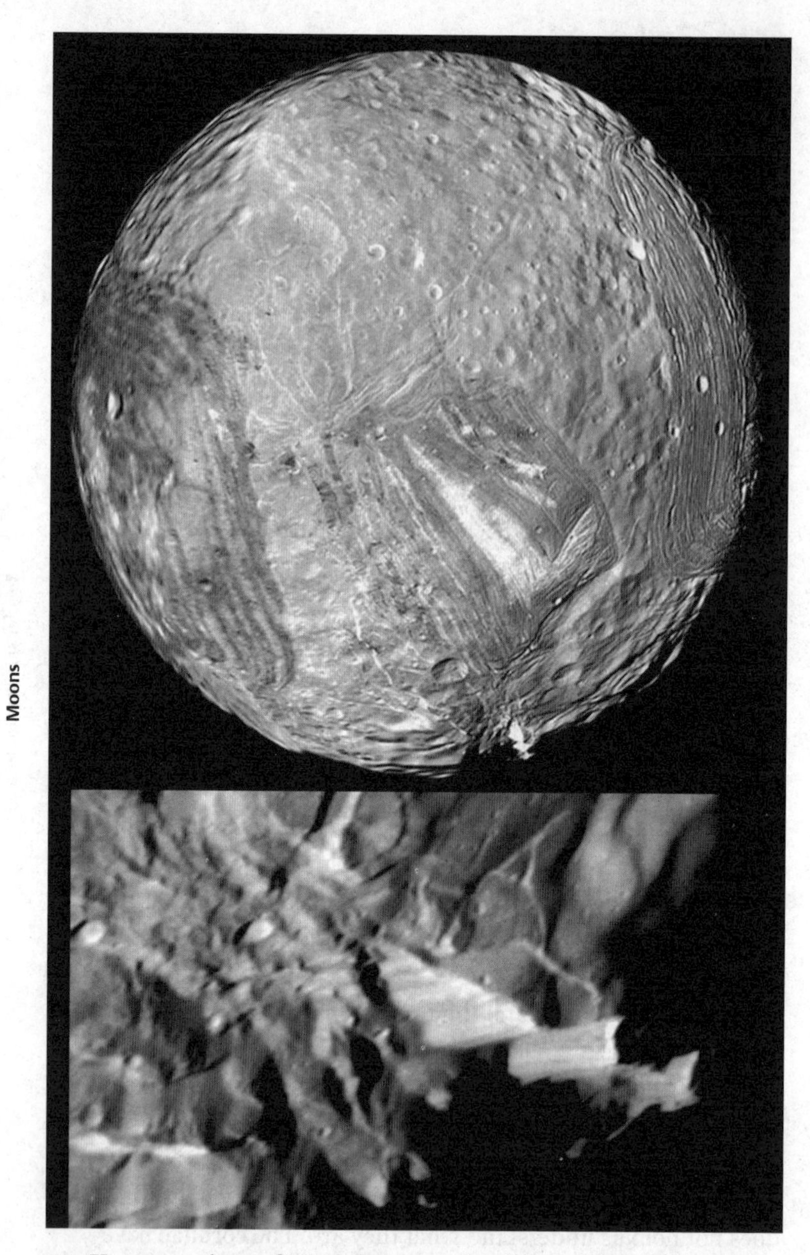

20. Voyager 2 views of Miranda. A mosaic of the sunlit hemisphere (top); slightly rotated view of the cliffs at the lower edge of the other image, seen at 0.7 km per pixel from a range of 36,250 km (bottom).

Inverness Corona, which cuts into the adjacent terrain and has produced some impressive cliffs up to 10 km high.

An early theory that each corona is a reaccreted fragment of a previous moon that had been broken apart by collision now seems too naïve, and we have to construct a more complex story. Miranda was probably once a passive, densely cratered globe, parts of whose surface still survive below the mantling deposit. It experienced one or more episodes of tidal heating, during which fracturing and localized cryovolcanism produced the coronae. The coronae may be surfaced by icy magmas that welled up through fissures to produce the ridges that dominate much of their area. If there were explosive eruptions at the same time, this could explain the muted profiles of the pre-existing impact craters on the terrain beyond the coronae.

This all makes sense, at least superficially, though it may turn out to be wrong when we eventually see Miranda, especially the as-yet unimaged half, in more detail (no mission to Uranus is currently planned). Tidal heating could have been driven by an interval of 3:1 orbital resonance with Umbriel, or 5:3 resonance with Ariel.

Ammonia is much more effective than salts at depressing the temperature at which a watery fluid will freeze, and given the likely abundance of ammonia in ices at this distance from the Sun, the cryomagma erupted on Ariel is probably an ammonia-water mixture that can be melted out from ice at a temperature as low as –97°C. This is lower than the melting temperature of either pure water or pure ammonia in isolation, and is an example of how mixtures of ices mimic the behaviour of mixtures of silicate minerals in rock, where also the melting temperature of rock is lower than the melting temperature of separate minerals.

Of Uranus' other regular satellites, Voyager 2 showed hints of major fractures on Oberon and obvious fractures on Titania, but only on Ariel did it reveal clear evidence of both fracturing and cryovolcanism.

Ariel

Ariel is the second largest of Uranus' satellites, and Voyager 2 saw it in moderate detail from a range of 127,000 km during its approach, before switching its attention to Miranda. It is denser than any moon of Saturn apart from Titan and is probably about half ice and half rock (which, this far from the Sun, could be quite rich in carbon). In addition to water ice, spectroscopic studies from Earth have detected carbon dioxide ice, but ammonia (which has no easily detected spectroscopic signature) is likely to be there in abundance too, to judge from the nature of the cryovolcanic resurfacing that is apparent in many areas (Figure 21).

Ariel's surface has many fractures, bounding down-dropped strips of terrain displaced by a series of fault movements at different times. This sort of faulting is symptomatic of stretching of the crust, and may reflect an episode of internal heating (which could have been tidal, or, if early enough, radioactive heating) causing thermal expansion of the interior. These valleys are widest in the centre and lower right of Figure 21, where it can be seen that their floors have fewer craters than the high ground to either side, showing that they have been flooded by something. This 'something' seems to have been very viscous, judging from what appears to be the edge of a flow cutting across the crater that is indicated by the arrow in Figure 21. On the Moon, where the lava was far less viscous, the flow would have spread across the entire crater and completely flooded it. An ammonia-water melt would move in a sufficiently viscous fashion under Ariel's low surface gravity (less than a tenth of the Moon's) and would fit the bill perfectly here.

However, a feature that seems not to fit with very viscous flow is the sinuous channel running near the middle of the wide faulted valley near the lower right of Figure 21. This is reminiscent of Hadley Rille

21. Voyager 2 view of Ariel, from about 130,000 km. The arrow marks an impact crater that has been partly flooded by a viscous cryovolcanic flow coming out of the faulted valley below and to the left. A slightly smaller adjacent crater is superimposed on this flow, and so must be younger.

(Figure 6) and many similar features on the Moon, whose origin is accepted as thermal erosion by fast-flowing lava or collapse of the roof of a lava tube—both of which require low viscosity.

There is so much we don't understand about Ariel, which has had a complex history, and which we are unlikely to sort out in my lifetime. How many different kinds of cryomagma were erupted on Ariel? Was there any sideways slip between the faulted blocks of terrain (there are hints of this if you look at the lower right of Figure 20)? What are those mountains silhouetted against the blackness of space at the upper left?

Triton

In many ways, Voyager 2 saved its best until last. It passed within 25,000 km of Triton, imaging about 40 per cent of the surface and revealing a complex surface that is rich in a diversity of ices, with a polar cap of nitrogen ice (Figure 22). I remarked on Triton's likely origin as a captured Kuiper belt object in Chapter 4. However although its composition is similar to Pluto, it differs markedly from Pluto in terms of landscapes and surface history.

Triton has a nitrogen atmosphere with traces of methane, but it is far less substantial than Titan's atmosphere and the surface pressure is only a few hundred thousandths of the Earth's. Tenuous though this is, it is sufficient to support a few wispy cirrus-like clouds of nitrogen ice particles a few kilometres above the surface and a 30 km high hydrocarbon haze layer. Triton has a global average albedo of 0.78, which is exceeded among regular satellites only by that of Enceladus. This is not so much because the surface is young (nearly 200 impact craters have been identified, all less than 30 km across), but because there is a frost of nitrogen that freezes out of the atmosphere to coat the –230°C surface.

Triton has peculiar seasons. Although its own spin is perpendicular to its orbit about Neptune, this orbit is inclined at 157° to the planet's equator (so it is retrograde). On top of this, Neptune's equator is tilted at 30° relative to its orbit about the Sun. A combination of these effects means that the sub-solar latitude of Triton varies between 55° N and 55° S, with a period somewhat longer than Neptune's 165-year orbit because Triton's orbital plane precesses as well, meaning that Triton experiences exceptionally long seasons. When Voyager 2 flew by, the southern hemisphere was part-way through its spring season, with the fringe of its polar cap receding as the nitrogen ice sublimed and added to the atmosphere. Presumably the northern polar cap, in darkness and unseen, was growing as northern autumn progressed towards winter and nitrogen froze out of the atmosphere and on to the cold ground.

Where visible beyond the polar cap and through the thinner nitrogen frost, the surface is a mixture of water ice and carbon dioxide ice, with traces of methane and carbon monoxide. Ammonia is suspected, particularly in view of the cryovolcanic nature of the landscape, but has not been detected.

Triton's terrain is crossed by a number of furrowed ridges, known as sulci (Latin for 'grooves'), which are reminiscent of the largest features on Europa such as the two that intersect near the top of Figure 15. These may sit above fractures that were opened and closed (or possibly slid sideways) under the influence of tidal forces before Triton's orbit became virtually circular. In the lower part of Figure 22, lengths of some sulci are buried by smooth plains that look as if they were emplaced as cryovolcanic lava flows. Smooth, presumed cryovolcanic plains occur also in two irregularly shaped depressions near the middle right of Figure 22.

The terrain in the upper part of the figure is older. Here the sulci are uninterrupted, except where a younger sulcus cuts an older one, and they cross terrain pockmarked by curious 30–40 km dimples. No one knows what caused these. This is called 'cantaloupe terrain', because of its resemblance to a melon skin, but that doesn't help much! The dimples could overlie places where individual pods of 'warm' ice have risen through the crust, but regardless of their exact nature they are surely another instance of cryovolcanic activity.

22. Voyager 2 mosaic using the best images of Triton. At the time of imaging, the fringe of the south polar cap was at 5–20° S. This view extends from about 60° S to 40° N.

Triton's cryovolcanism and the stressing of the crust demonstrated by the sulci were presumably powered tidally, during the perhaps billion years that it took for Triton's orbit to become circular after its capture by Neptune. We don't know how long ago that was, but the impact craters scattered across the surface suggest that this activity has now ceased, or at least waned.

However, there are still eruptions of a kind on Triton. These occur through the polar cap, and were seen in progress (though imaged rather poorly) by Voyager 2. What seems to happen here is that sunlight passes through the largely translucent nitrogen ice and warms the darker substrate below. This warmth is conducted up into the nitrogen ice, until its base begins to sublime. A blister of nitrogen gas swells up beneath the icecap until it bursts, allowing the gas to jet skywards, taking with it dusty grains from the dark substrate. The plume becomes neutrally buoyant at a height of several kilometres. It is then blown downwind, and the dark particles begin to fall out, leaving low-albedo streaks that can be seen on the polar cap. If this explanation is correct, these are solar-powered geysers rather than true cryovolcanic eruptions. Nevertheless, they qualify Triton for membership of the exclusive club of moons with proven present-day activity that otherwise includes only Io, Titan, and Enceladus.

Triton is a wonderful world, but it won't be around for ever. Its orbit is already closer to Neptune than the Moon is to the Earth, and tidal processes to do with its inclined orbit passing over the planet's equatorial bulge are gradually dragging it closer. Calculations suggest that Triton will pass within Neptune's Roche limit in about three billion years, which could give birth to a ring system far more spectacular than Saturn's.

Future missions

It should be clear that there is much more to find out about even the best-known moons of the giant planets, and that more missions will be needed to deliver this knowledge. The earliest mission currently approved by any space agency is the ESA's Jupiter Icy moons Explorer (JUICE), planned for launch in 2022 and arrival for a thirty-month tour at Jupiter in 2030. This will begin with a series of close fly-bys of Europa and Callisto, using assists from their gravity until eventually the probe can be captured into orbit about Ganymede, into which it will eventually crash. Also for launch sometime in the 2020s is NASA's unimaginatively named Europa Mission. Like JUICE, this will orbit Jupiter but make dozens of flybys of Europa to study its subsurface ocean and to look for eruption plumes. A subsequent NASA or ESA mission, not yet funded, might usefully drop a series of penetrators or tiny 'chipsats' into the ice beside young cracks in Europa, to probe the ice and search for life.

Titan is an appealing target as well—and proposals for splashdown into one of Titan's largest methane seas are under consideration at NASA. I hope that such a mission would also get a chance to analyse Enceladus plume material for biological signatures. New missions to Triton and the moons of Uranus, of whose surfaces we have tantalizingly seen less than half, are lower down most priority lists.

Chapter 6
The moons of Mars: captured asteroids

Mars is the planet most recently found to have moons. In his fictional satire of 1726, *Gulliver's Travels*, Jonathan Swift described the astronomers of the flying island of Laputa as having discovered two small moons of Mars: 'They have likewise discovered two lesser stars, or satellites, which revolve about Mars; whereof the innermost... revolves in the space of ten hours, and the latter [outermost] in twenty-one and a half' (Part III, Chapter 3).

In reality it was a century and a half later, in 1877, that the moons of Mars were actually discovered, two in number as Swift has guessed, by the American Asaph Hall after a protracted search using what was then the world's largest refracting telescope (66 cm). He found the smaller, outer one (Deimos) first, and the larger, inner one (Phobos) six days later. Their actual orbital periods (see Appendix) are 7.7 hours and 30.2 hours respectively.

Swift's guess that Mars has two moons was probably an attempt to fit Mars into a pattern, bearing in mind that the Earth has one moon and (at the time) Jupiter had four known moons and Saturn five. Kepler made a similar numerical speculation in 1610, inspired by Jupiter's four moons before any of Saturn's had been discovered. The short orbital periods that Swift gave to his invented moons indicates his awareness that if any moons of Mars had longer periods, they would be far enough from the planet to have been discovered already by astronomers in the real world.

Both of them are small rocky bodies in synchronous rotation. Phobos is about the size of Saturn's Trojan moon Calypso (Figure 10) and Adrastea, the smallest inner moonlet of a giant planet included in Figure 12. Phobos measures 27 × 22 × 18 km, and Deimos 15 × 12 × 11 km. They are far too small for their own gravity to pull their shapes into hydrostatic equilibrium. Despite the fact that their orbits are prograde, only slightly eccentric, and inclined at only about 1° to Mars' orbital plane, they are almost certainly captured asteroids. This makes them more analogous to the irregular moons of the giant planets than to their inner moonlets. They have little in common with the Moon.

Their densities, just less than twice that of water, are too low for them to be solid rock. There may be some ice inside, though spectroscopic studies show no signs of hydration at the surface. It is more likely that their low densities, a property they have in common with most of those asteroids whose densities have been determined, are because below the surface regolith their interiors consist of chunks (unknown in size) of loosely packed rubble. This is a similar explanation to that offered for Saturn's Hyperion, an icy moon whose density is too low to be solid ice. Spectroscopically, both Phobos and Deimos resemble asteroids that are believed to equate to a class of meteorite known as carbonaceous chondrites.

Phobos

Phobos (Figure 23) has been imaged in better detail than Deimos, because its orbital height of only 5,645 km above the surface (as opposed to Deimos' 20,000 km altitude) brings it closer to spacecraft orbiting Mars, of which there have been several at typical altitudes of about 300 km. The Russian space agency has attempted three missions to Phobos. Of the duplicates Phobos 1 and 2, launched in 1988, only Phobos 2 made it to Mars, but contact was lost in January 1989 as it closed in on Phobos itself to attempt a landing. More recently, Phobos-Grunt, which was intended to bring back a sample from the surface, malfunctioned and failed to leave Earth orbit after launch in 2011.

Phobos has numerous craters. The younger ones are sharp but older examples are very muted, as if buried by regolith. Using the first close-up images of Phobos from the Mariner 9 Mars orbiter, seven craters were named in 1973: six after astronomers who had worked on Mars, including Hall himself. The seventh and largest (9 km in diameter) was named Stickney, after Angeline Stickney, Hall's wife, to whose encouragement, when he had been on the point of giving up his search, he attributed his eventual success in finding Mars' moons. Stickney occupies almost the whole of the lower-left end of Phobos in Figure 23, and is not far from the middle of Phobos' leading hemisphere. Although large in comparison to Phobos, Stickney is scarcely bigger than the lunar crater Copernicus B in Figure 4. It is far too small to have developed a central peak, and has a simple bowl shape. There is a prominent 2 km crater within it named Limtoc, one of eight names assigned in 2006 that were taken from *Gulliver's Travels* (Limtoc being the name of a Lilliputian general).

Individual craters on Phobos are surely the results of impacts, but there is a great deal of controversy about the 100–200-metre-wide grooves that can be seen crossing the surface. In detail, some are simple grooves whereas others are chains of overlapping pits. Many grooves appear to emanate (or radiate) from Stickney, though a few cross Stickney's rim and so must be younger than this crater. Families of grooves of different orientations can be recognized (especially beyond the area shown in Figure 23), with consistent age relationships exhibited by the overprinting of older-family grooves by younger-family grooves.

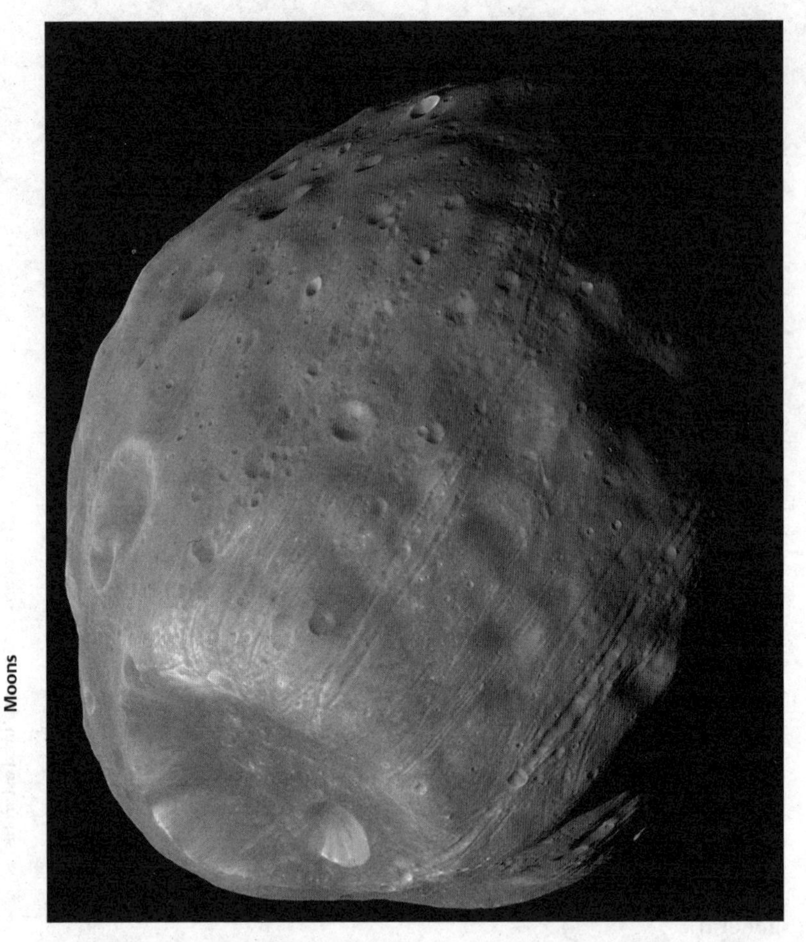

23. The Mars-facing side of Phobos, recorded by the HiRISE camera of NASA's Mars Reconnaissance Orbiter with a resolution of about six metres per pixel from a range of 5,800 km.

Early hypotheses to explain the grooves tried to relate them to Stickney. Could they be chains of overlapping secondary craters made by ejecta flung out from Stickney? This doesn't work, because any ejecta from Stickney travelling fast enough to make such craters would escape from Phobos' low gravity and so could not come back to strike the surface, and in any case some grooves are too young. Could they be fractures caused by the

Stickney impact? This falls down because of the various ages and orientations of groove families.

The near-radial disposition of most grooves with respect to Stickney is now mostly dismissed as a red herring, with Stickney being a coincidental impact near Phobos' leading point. If that is correct, the relationship that actually needs explaining is the one between the groove pattern and Phobos' orbital travel. There are two competing hypotheses based on this. One has the grooves as the surface manifestations of fractures that were imposed on Phobos' interior by tidal forces, or by aerodynamic drag from an extended ancient Martian atmosphere during capture of Phobos by Mars. Sets of fractures of the observed orientations could be formed in this way, and later widened to form the grooves. This is plausible, but the exactly planar fractures implied by the straightness of the grooves is hard to reconcile with the likely rubbly and porous interior required by Phobos' low density.

The other motion-based hypothesis notes that Phobos' low orbit must inevitably from time to time carry it through a hail of ejecta flung out by large crater-forming impacts on Mars. Fragments of this ejecta could hit Phobos either while still travelling upwards or when falling back down, which would account for the grooves on the side of Phobos that never faces the planet. Each groove would represent Phobos' passage through a string of ejecta, leaving a trail of damage on its surface like that on the bodywork of a car that had been driven through the fire of a machine gun. Several strings of ejecta from the same impact on Mars would be encountered at once, which is why the grooves come in families. There are objections to this explanation too, notably the remarkable lack of dispersion of the ejecta travelling from Mars required to make such straight grooves (i.e. each component of a string of ejecta would have had to travel in *precisely* the same direction, with no sideways deviation), so perhaps it is best to regard the grooves of Phobos as a mystery that has yet to be adequately explained.

Deimos

Deimos (Figure 24) has no grooves, probably not so much because it lacks any crater of the size of Stickney, but because it is further from Mars. This makes it less vulnerable to tidal or capture-related stresses and also less likely to encounter intense hails of ejecta from the Martian surface.

Deimos has a cratered surface similar to the smoother parts of Phobos. Only two of its craters have been named (both in 1973, based on Mariner 9 images): Swift (1 km in diameter, identified by the converging arrows in Figure 24) and Voltaire (1.9 km in diameter), which is a more subtle, less prominent feature immediately below Swift in Figure 24. The crater Voltaire is named after the French Enlightenment writer who, a quarter of a century after Swift, set down his own opinion that Mars must have two moons.

Moons in the Martian sky

Phobos and Deimos are the only moons in our Solar System (apart from our own) that could one day be admired in the sky by humans standing on the surface of the relevant planet, because Mars is the only planet with moons which has a surface that could be landed on. Being so much closer to Mars than the Moon is to the Earth, and in low-inclination orbits, they pass into Mars' shadow at night more often than lunar eclipses occur on Earth. Similarly, by day they often cross the disc of the Sun, but despite being close to Mars their size is too small to hide the whole of the Sun's disc, so there are no total solar eclipses on Mars.

When overhead, Phobos would look almost half the size of the Moon in the Earth's sky, but when close to the horizon it would be nearly twice as far away from the observer and so would look proportionately smaller. This effect is not noticeable for the Moon, whose orbit is much bigger than the Earth's radius.

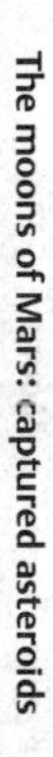

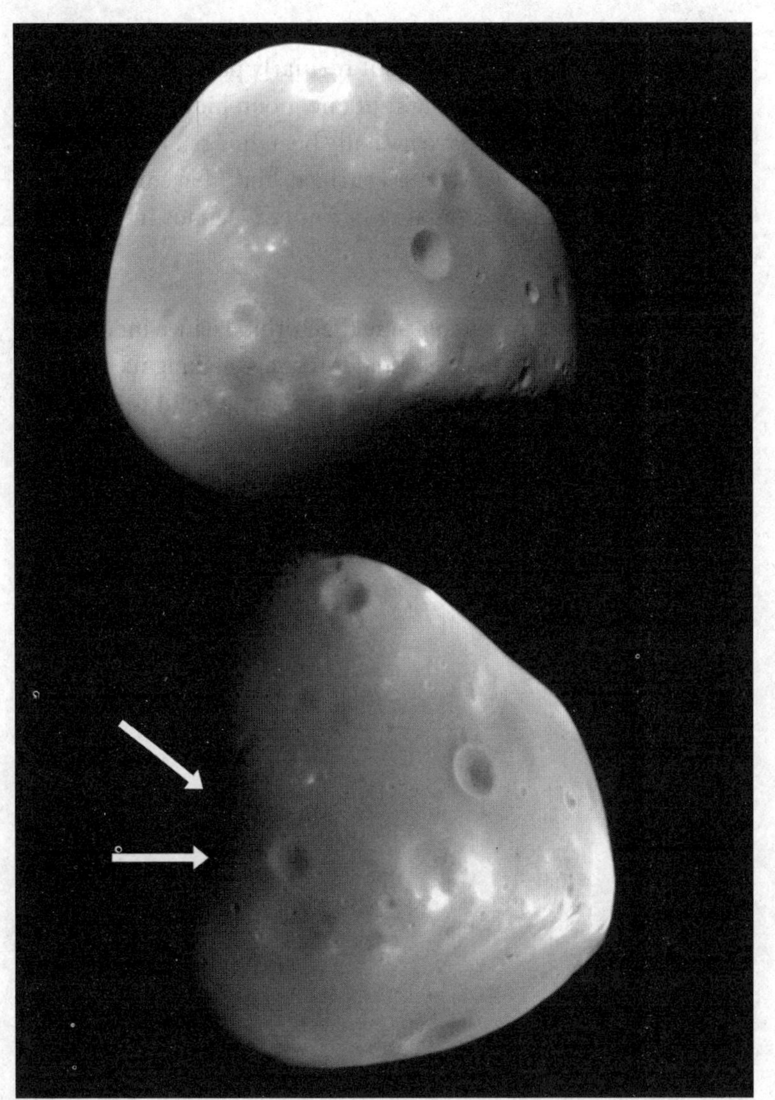

24. Two views of the same (Mars-facing) face of Deimos under different illumination, recorded by the HiRISE camera of NASA's Mars Reconnaissance Orbiter with a resolution of about twenty metres per pixel. The crater Swift is indicated by the converging arrows.

Phobos travels across the Martian sky faster than Deimos, and seen from low latitudes on Mars it regularly passes in front of it, presenting the remarkable sight of one moon hiding another. All of these phenomena have actually been imaged by cameras carried by rovers on the Martian surface, and I have included a link to a video of Phobos passing in front of Deimos in the Further reading.

Phobos and Deimos won't continue their orbital dance indefinitely. Phobos is so close to Mars that tidal forces are making its orbit decay much faster even than Triton's. Its nearest point to Mars is getting lower by about two centimetres each year, and at that rate it can survive for about only fifty million years more before being ripped apart or crashing to the surface.

Chapter 7
Moons of small bodies

The small bodies of our Solar System can be summed up as: asteroids (rocky or carbonaceous objects that are concentrated in, but not confined to, the space between the orbits of Mars and Jupiter); trans-Neptunian objects (icy bodies beyond Neptune's orbit, including Pluto); and comets (small icy bodies with strongly elliptical orbits that can come close to the Sun). These three types are useful distinctions, but not all objects can be neatly pigeonholed, as you will see shortly when a fourth type, centaurs, comes into the story.

Among all these, only comets are devoid of known moons. Comets have been known since antiquity only because they can develop spectacular tails of gas and dust when they pass close to the Sun. The solid part, or 'nucleus', of the largest comets is much smaller than the largest examples of any other kind of small body. Most comet nuclei are less than 10 km across, and there is no known comet nucleus as large as 100 km.

Three comet nuclei have been shown to have shapes consistent with being 'contact binaries', that is two main lumps loosely held together beneath a regolith layer, but no comet has been found to have a freely orbiting moon. Indeed, if a comet did have a moon it would be unlikely to survive for long, because vented gas would make their mutual orbits chaotic.

Asteroids with moons

Well-characterized asteroids are formally designated by a name preceded by a number corresponding roughly to the sequence of discovery. The first asteroid to be found was 1 Ceres (in 1801), which is also the largest at 950 km in diameter. Those that have not been tracked long enough to be sure of their orbit are unnamed, and have only a provisional designation consisting of the year of discovery followed by a two-letter (or two-letter-plus-number) code. Asteroids as small as about ten metres in size can be tracked by radar when they pass close to the Earth, but may never be seen again.

In 2015, 184 asteroids were known to have moons. Of these, 104 are in the main asteroid belt between the orbits of Mars and Jupiter, 21 pass inside Mars' orbit, and 55 cross or come close to Earth's orbit. Most of them have only one moon, but nine have two moons.

This remarkable tally is thanks mainly to advances in optical telescope imaging over the past twenty years. However, the first asteroid moon was discovered by accident in February 1994 when images from Galileo's fly-by of the asteroid 243 Ida six months previously were downloaded. These revealed a 1.5 km diameter moon that was later named Dactyl, whose cratered surface demonstrates that it is fairly old. It has a prograde orbit about Ida and apparently similar composition (Figure 25).

Dactyl remains the only moon of an asteroid to have been seen at close quarters, and to explain it we need to consider Ida's history. Ida is a member of the Koronis family of main-belt asteroids, which share similar orbits and are thought to result from a collision between two larger bodies about two billion years ago. There are twenty Koronis family asteroids like Ida that exceed 20 km in diameter and nearly 300 known smaller ones. Dactyl

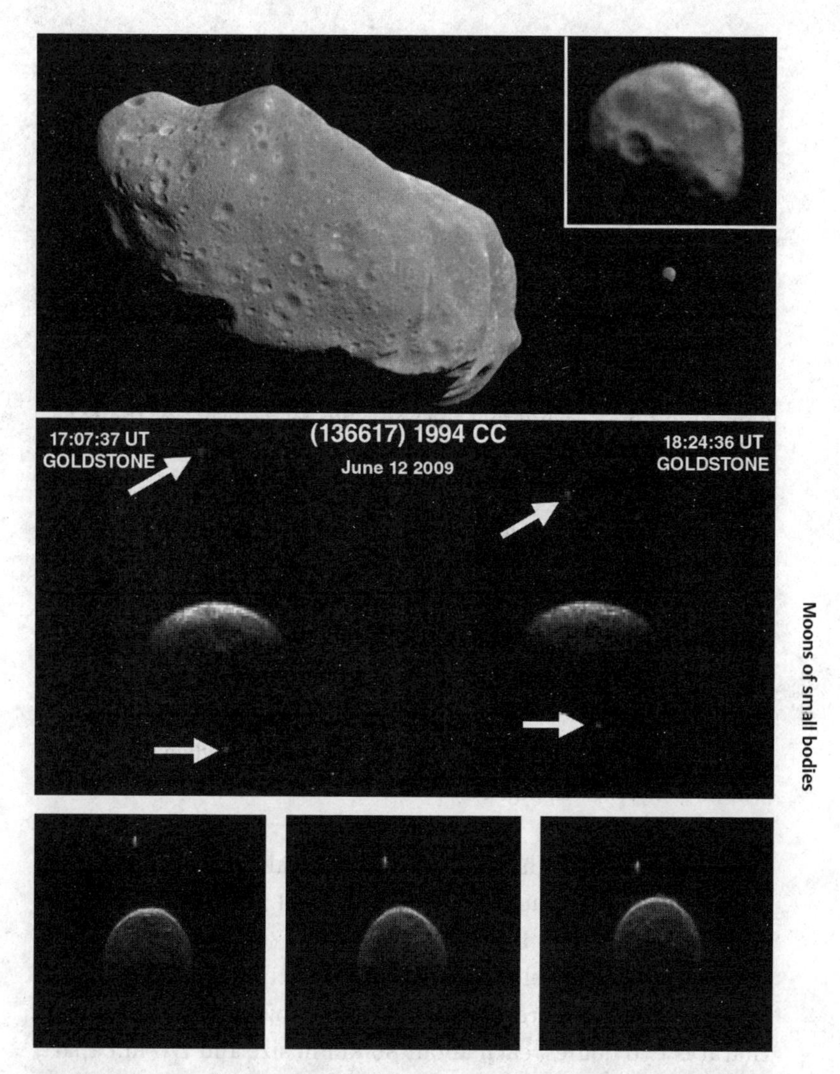

25. Asteroid Ida (56 km long) and its 1.5 km moon Dactyl in a single Galileo image (top), the inset is the highest-resolution view of Dactyl, enlarged, seen from a range of 3,900 km; views of 700-metre asteroid 1994 CC and its two tiny moons (arrowed), imaged by radar at a range of about 3,000,000 km in 2009 (middle); three successive radar views of asteroid 2004 BL_{86} and its seventy-metre moon, at a range of only 1.2 km from Earth on 26 January 2015 (bottom).

may be a fragment from the same parent body that broke away from Ida during this collision, but with too little force to become independent, or it could be a chunk knocked off Ida more recently.

All the other known moons of main-belt asteroids have been discovered telescopically, beginning in 1998 with Petit-Prince, a 13 km diameter moon taking 4.8 days to orbit the 213 km diameter asteroid 45 Eugenia at a distance of 1,184 km, close to its equatorial plane. Discovery of a 6 km diameter, closer-in, second moon of 45 Eugenia was announced in 2007. This has yet to receive an official name, but I understand that the name 'Princesse' is before the IAU for approval.

Before the second moon of 45 Eugenia had been discovered, the 286 km asteroid 87 Sylvia had already been found to have two moons, about 18 km and 7 km across and orbiting very close to Sylvia's equatorial plane at distances of 707 km and 1,357 km. These have been given the highly appropriate names Romulus and Remus (after the wolf-suckled twins in the story of Rome's founding) whose mythological mother was Rhea Sylvia after whom 87 Sylvia was named.

The examples so far have all been of a sizeable asteroid with either one or two much smaller moons, but 90 Antiope is something different. It was discovered in 1866, and seemed to be an unremarkable largish asteroid until the year 2000 when it began to be revealed as a binary object. We now know that it is two bodies, each about 86 km in size and 170 km apart, in synchronous rotation and orbiting their common centre of mass (or 'barycentre') with a period of 16.5 hours. Being indistinguishable in size or mass, each can be regarded as a moon of the other.

Much smaller asteroids can be seen only when they come close to the Earth, and it is even possible to study them by radar when less than about ten million kilometres away. Of the 'near-Earth asteroids', 16 per cent have been shown to be double or triple systems. For example, 69230 Hermes consists of two nearly identical objects about 400 metres across and only 1,200 metres apart, making them a miniature version of 90 Antiope. Other examples look like scaled-down versions of Ida/Dactyl or Sylvia/Romulus/Remus. Among these, the 2.8 km diameter 1998 QE_2 has been shown to rotate in about five hours, and its 600-metre-wide moon is in synchronous rotation in a thirty-two-hour orbit at a range of about 6 km. The asteroid 1994 CC (Figure 25) has two even tinier moons. The inner is about 110 metres across and seems to be in synchronous rotation in a 1.7 km radius orbit taking about thirty hours, whereas the outer (about eighty metres in diameter) seems to rotate faster than its nine-day orbital period. This was the smallest known moon of any object until the 355-metre asteroid 2004 BL_{86} passed Earth at a range of 1.2 million km in January 2015 and radar showed a moon only seventy metres across (Figure 25).

Many asteroid moons probably originated as debris knocked off the main body by an impact, but in the case of the smaller asteroids their moons could have begun as loosely held components flung off the main body by a centrifugal process as rotation speed increased. This surprising spin-up can happen because of something known as the YORP (Yarkovsky–O'Keefe–Radzievskii–Paddack) effect, whereby the angular momentum of a small, irregular rotating body is unbalanced because of absorption of sunlight on the dayside only, but emission of thermal radiation from the warmed surface that extends into the nightside.

Centaurs with moons

Centaurs are asteroid-like bodies of dominantly icy rather than rocky composition, whose orbits lie beyond that of Jupiter but inside Neptune's. About 200 are known, of which the largest example is smaller than the largest Kuiper-belt objects that are found beyond Neptune.

Moons have been imaged telescopically for two objects that are sometimes classified as centaurs: these are 42355 Typhon (162 km) and its moon Echidna (90 km), and 65489 Ceto (200 km) and its moon Phorcys (170 km). However, although these come inside the orbit of Uranus when at their closest to the Sun, their orbits are strongly elliptical and take them way beyond Neptune when at their most distant, so they are perhaps more appropriately regarded as inwardly scattered trans-Neptunian objects.

For an example of an uncontested centaur that may have moons, we need to turn to 10199 Chariklo. This is the largest known centaur, and has a diameter of about 250 km. In June 2013 it was predicted to pass exactly in front of a star (a rare event for an object so small) as seen from various South American observatories. Exact measurement of the interlude (typically a few seconds long) when the star is hidden, known as an occultation, provides the best way we have to work out the size of such a small distant object, so every telescope in the path of the occultation (which had the same width as Chariklo itself) was looking. To everyone's surprise, the star was dimmed briefly twice before and again twice after the main event. The only feasible explanation for this is that Chariklo has two rings: a 7 km wide ring at 391 km radius and a less dense 3 km wide ring at 405 km.

Chariklo is by far the smallest body found to have rings. These rings are very narrow, so unless they formed less than a few thousand years ago (which would be a remarkable coincidence) they are probably confined by the presence of kilometre-size, unseen shepherd moons.

Trans-Neptunian objects and their moons

When Pluto's largest moon was discovered, in 1978, Pluto was still regarded as a planet, as it had been ever since its discovery in 1930. However, Pluto was always a misfit, because it passes temporarily inside the orbit of Neptune, which is 10,000 times more massive. The two are in 2:3 orbital resonance about the Sun (Pluto orbits twice in the time it takes Neptune to complete three orbits), but they can never collide because whenever Pluto is at its closest to the Sun, Neptune is either about 50° ahead or about 50° behind. Moreover, Pluto's orbit is inclined in such a way that it is well above the orbit of Neptune when closest to the Sun, and in fact the distance between Pluto and Neptune always exceeds seventeen times the Earth–Sun distance.

In 1992, a second Pluto-like trans-Neptunian object was discovered, eventually named 136199 Eris, and now believed to be about 27 per cent more massive than Pluto although fractionally smaller in size. By 2015 the tally of trans-Neptunian objects exceeded 1,500, of which about eighty (including Pluto and Eris) are known to have moons. Many of these are in Pluto-like orbits in a band thirty to fifty times further from the Sun than the Earth's orbit. This region is referred to as the Kuiper belt, after Gerard Kuiper (1905–73) who predicted something similar. Beyond lies the 'scattered disc' of similar objects in more eccentric, and often more steeply inclined, orbits that take them out beyond a hundred Earth–Sun distances.

Pluto is one of the largest, but is not exceptional and is not distinct from the continuum of trans-Neptunian objects other than having been the first to be discovered, by some sixty years. This new knowledge of the Solar System made it illogical to continue to regard Pluto as a planet, unless Eris and several others were also to be classified as planets. At its annual meeting in Prague in 2006 the IAU voted to adopt a definition of planet that excluded bodies that do not greatly out-mass objects sharing similar orbits and that cross the orbit of a much more massive object. This demoted Pluto while leaving the status of the other eight planets unchanged.

This decision was (and still is) unpopular in some quarters, but was in my view the right one, and much less messy than any alternative. Those objects orbiting the Sun that do not qualify as planets but that are massive enough for their gravity to pull their shapes into hydrostatic equilibrium are classified as dwarf planets. This class includes one rocky asteroid (1 Ceres), and at least five icy trans-Neptunian objects of which Pluto is one. This definition sounds logical, but in practice it cannot always be applied with confidence, because only 1 Ceres and Pluto have been seen clearly enough to demonstrate their shapes. The others are merely assumed to have shapes in hydrostatic equilibrium on the basis of their estimated size and mass.

When Pluto's first moon was discovered, by the American James Christy (1938–) in 1978, using a 1.55-metre telescope, it was detectable merely as a bulge that was sometimes present on the side of blurred images of Pluto, though newer telescopes and modern techniques soon enabled the two bodies to be resolved. Christy proposed the name Charon for this moon, and this was accepted by the IAU. Christy, it is said, would have liked to name his discovery after his wife, Charlene, known as 'Char'. He knew that the IAU would not accept this, but realized that Charon would seem appropriate because in Greek mythology Charon was the ferryman who conveyed the souls of the dead across the river Styx to Hades, of which Pluto (the Roman name) was the ruler. The orthodox way to pronounce the ferryman's name is 'Kairon' but Christy, and most Americans, say 'Shairon' or 'Sharon', preserving the 'sh' sound of Charlene.

Pluto is now known to have five moons (see Appendix), of which Charon is by far the largest. Its mass is one-eighth that of Pluto, so the two bodies are much more similar than the Earth and Moon, where the mass ratio is 1:80. Because of this, their barycentre is not inside Pluto but occurs about one Pluto radius above its surface, in the direction of Charon. The two bodies mutually orbit this point, and are tidally locked into synchronous rotation so that there is one hemisphere of Pluto from which Charon can never be seen, as well as one hemisphere of Charon from which Pluto can never be seen. There are also, of course, longitudes on Pluto experiencing eternal moonrise or moonset.

Pluto's axis is inclined at 119.6° to its orbit (so, like Uranus', its rotation is retrograde). All its moons' orbits lie within a fraction of a degree of its equatorial plane, and are prograde (travelling in the same direction as Pluto's spin) and nearly circular. The orbital periods from inner (Charon) to outer (Hydra) are nearly (but not quite) in 1:3:4:5:6 resonance.

The names of the smaller moons follow the underworld theme begun by Pluto and Charon: Nix, the goddess of darkness and the night, was the mother of Charon (the ferryman, not Mrs Christy); Hydra was a nine-headed serpent, and some say that this is a reminder of Pluto's temporary reign as a ninth planet; Styx is named after the goddess of the river across which Charon the ferryman plied his trade; and Kerberos was the three-headed dog that guarded the entrance to the underworld. Similar underworld names were in reserve for any other moons discovered by the New Horizons probe that flew past in July 2015, but it did not find any. However, during an all-too-brief encounter at a speed of 14 km/s, passing about 10,000 km from Pluto and 27,000 km from Charon, it revealed large regions of their globes in glorious detail, with a smallest pixel size less than 200 m.

Charon may have formed as the result of a 'giant impact' event similar to that proposed for the Moon's origin, in which case the four smaller moons could be debris from the same event that has migrated outwards. If Nyx and Hydra turn out to be in synchronous rotation this will suggest that outward migration has occurred, because calculations show that tidal forces would be inadequate to achieve this at their current distances. If they are not in synchronous rotation, this will suggest capture (with all its attendant difficulties) as a more likely origin.

Pluto is covered by ices of nitrogen, methane, and carbon monoxide, overlying stronger and less-volatile 'bedrock' made of water-ice, but is dense enough that it must have rock in its deep interior. It has an atmosphere derived from its surface ices, dominated by nitrogen. In all these respects it appears to resemble Neptune's large moon Triton, which you may recall is likely to have been captured from the Kuiper belt. New Horizons close-up images found Pluto's surface to be even less-cratered than Triton's, showing that it has been resurfaced even more recently. Some high resolution frames seem to contain no craters at all, suggesting a local surface age of less than 100 million years. Triton-like sulci and cantaloupe terrain are absent. Instead, there are tracts where water-ice sticks up through the other ices to form jagged 3 km high mountains, and plains of carbon monoxide ice whose surface looks like a larger scale version of ground patterned by freeze-thaw process in the Earth's arctic and parts of Mars.

In contrast, Charon's surface is dominated by water ice, but hydrated ammonia was detected telescopically even before New Horizons, prompting speculation of recently active cryovolcanism. The amount of tidal heating in the Pluto–Charon system is not well understood, but the varying effect of the Sun's pull on the steeply inclined Pluto–Charon orbit, to which Charon is more vulnerable than Pluto, might be enough to generate the heat required to partially melt ice mixtures in their interiors. First thoughts from the New Horizons team were that the amount of recent heating required to account for Pluto's terrain could not be explained tidally. Although they could cite no viable alternative mechanism, they speculated that this cast doubt upon the tidal heating explanation for young surfaces on icy bodies elsewhere too. A simpler, but equally contentious explanation is that Charon was captured, or formed by a giant impact, in the relatively recent past rather than billions of years ago.

Charon too has a youngish surface (Figure 26), though there are fewer signs of recent activity than on Pluto. It is notable for faulted valleys in the form of straight-sided troughs a few kilometres deep similar to those seen on Uranus's moons Ariel (Figure 21) and Titania, and a dark north polar cap that the New Horizons team named 'Mordor Regio'.

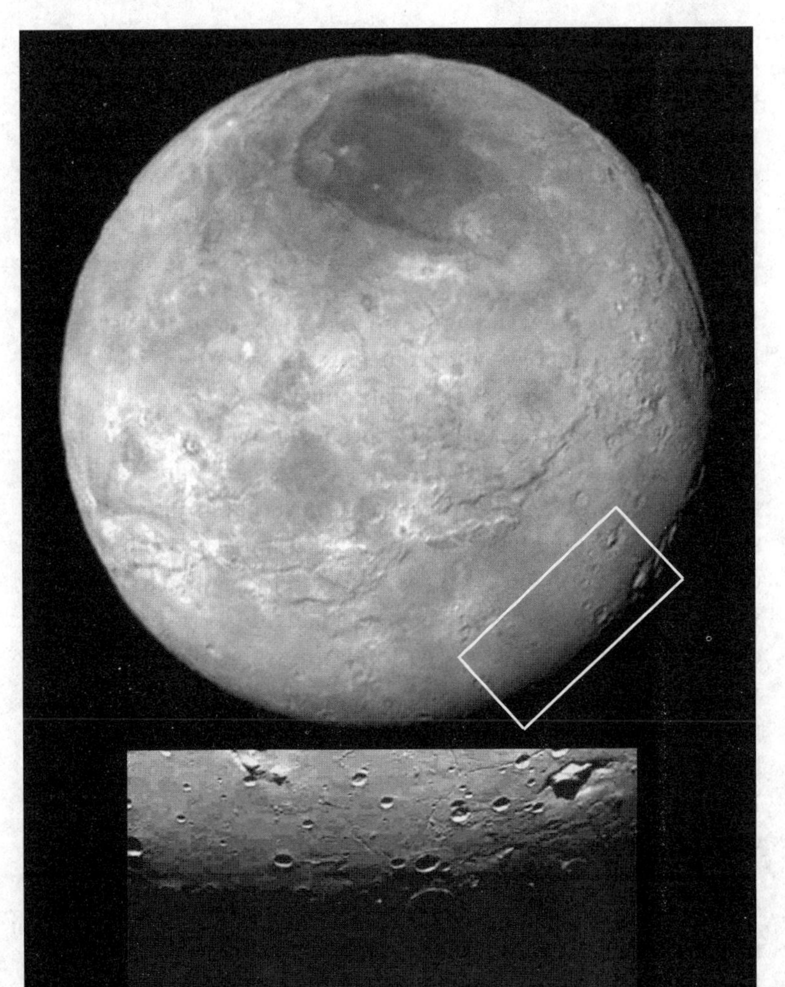

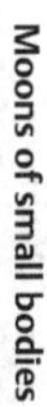

26. A global view of Charon by New Horizons. Note the north–south fracture system close to the eastern limb (right hand edge) and a second major fracture system running close to the equator. The dark polar region is Mordor Regio. Impact craters near the terminator are revealed by their interior shadows. Elsewhere they show up because of surrounding bright ejecta deposits, or sometimes because they have excavated darker material. The box locates the area shown in a more detailed image below (recorded several hours later so the terminator has migrated). This shows a perplexing mountain peak surrounded by a moat, and at the limit of resolution several straight fractures reminiscent of those west of Jansen E in Figure 5. There they are extensional fractures in mare basalt attributed to shallow intrusion on a vertical curtain of magma (a dyke), on Charon they could be caused by similar intrusion of a cryomagma made of a water-ammonia mixture.

As well as mapping the landforms and compositions of Pluto and Charon, New Horizons looked back at the Sun and the Earth as they were occulted in turn by each body to learn about Pluto's atmosphere and to try to determine whether Charon has any atmosphere at all. The smaller moons were imaged well enough to reveal their shapes, which are irregular as expected.

In 2018 or 2019 New Horizons will make a fly-by of at least one other Kuiper-belt object and its moons (if any). Moons of trans-Neptunian objects in general come in many combinations, and it is likely that most owe their origins to collisions. The object 136199 Eris bears the name of the ancient Greek goddess of strife and discord, a reference to the hornet's nest that its discovery stirred up regarding Pluto's status. It has a single known moon discovered in 2005 that has been allocated the name Dysnomia, which was borne by Eris' mythological daughter. Eris has a radius of about 1,200 km, whereas Dysnomia's must be somewhere in the region of 100–300 km, depending on its albedo. Dysnomia's orbit proves that the mass of Eris is significantly (about 27 per cent) larger than Pluto's, although its size is slightly smaller. Eris and Dysnomia are challenging objects to observe, because they are currently three times further from the Sun than Pluto is.

Size estimates of such distant objects are very uncertain, because they are mostly too small and too far away to show resolvable, measurable discs in a telescope. If a stellar occultation is observed and its duration measured, you can use the body's orbital speed to deduce the size of the part of the body that passed in front of the star, though that is likely to be a chord rather than a full diameter. These are rare events, which is why observatories were geared up to observe the stellar occultation by 10199 Chariklo that fortuitously revealed its rings. In the absence of any occultation data, if you assume an albedo you can use an object's brightness to deduce size. The uncertainties in this method meant that we were for a while unsure whether Eris was larger or smaller than Pluto, even though we were already sure that Eris was more massive.

Pluto and Charon's respective sizes were initially deduced in a series of mutual occultations that occurred between 1985 and 1990 when the plane of their orbits crossed the line of sight from Earth. A similar series of events between July 2014 and October 2018 should reveal the true sizes of the approximately 80 km Kuiper-belt object 385446 Manwë and its approximately 50 km moon Thorondor (whose names derive from the works of J. R. R. Tolkien). This is of more than passing interest because the current best estimate of Manwë's size means that its density is less than that of water, in which case it must be an icy rubble pile. Manwë's orbit round the Sun is in 4:7 resonance with Neptune's, as opposed to Pluto's 2:3 resonance.

One of the largest Kuiper-belt objects is 136108 Haumea , with a mean radius estimated at about 650 km. Its brightness goes up and down in a little less than two hours, revealing it as an elongated body with a 3.9-hour rotation period. The remarkably fast spin is believed to be responsible for distorting it into an elongated ellipsoid rather than merely bulging at the equator, like a slower-spinning body would. Its pole-to-pole diameter is probably just under 1,000 km, whereas the maximum and minimum diameters across its equator are about 1,960 and 1,520 km. It is dense enough to have a rocky interior, but its surface spectrum is that of crystalline water ice.

Haumea has two known moons, Hi'iaka (340 km diameter, orbitting at 49,900 km in 49.5 days) and Namaka (170 km diameter, orbitting at 25,700 km in 18.3 days). Like Haumea itself, these have the spectroscopic properties of crystalline water ice and a likely high albedo. Because of their shared properties, origin by capture is improbable and Haumea's moons are probably fragments that were flung off during an interlude of even more rapid spin or from a collision. Haumea is named after of a matron goddess of Hawai'i, whose many children (including Hi'iaka and Namaka) sprang from various parts of her body, as indeed the moons of the same names may have done.

The only other trans-Neptunian triple system apart from Haumea occurs in a Pluto-like orbit (2:3 resonance with Neptune) and is known as (47171) 1999 TC_{36}, not yet having been formally named. It consists of a central binary object, with each component about 260 km in diameter, so we are in the same dilemma as for Antiope as to which is the moon of which. However, in this case the binary pair is orbited by a 140 km moon. The average density of the two binary objects, estimated without the aid of mutual occultations to confirm their sizes, appears to be only about two-thirds that of ice. If this is correct, then each is probably an icy rubble pile.

Chapter 8
Moons in other planetary systems: exomoons

The first definite discovery of a planet around another star (an 'exoplanet') was made in 1995. Twenty years on, we know of more than 1,000 stars with exoplanets, of which nearly half have more than one. It now seems likely that 20 per cent of Sun-like stars have at least one giant exoplanet whereas at least 40 per cent may have lower-mass exoplanets.

In our Solar System, moons are considerably more numerous than planets, and it would be surprising if exomoons did not outnumber exoplanets. However, they will be very challenging to detect. Only a few exceptional exoplanets have been seen by direct imaging, and any exomoons are presently well below the visibility threshold. The vast majority of exoplanets are inferred either by cyclical changes in their star's radial velocity as it orbits the barycentre between itself and its exoplanet, or, in cases where the exoplanet's orbital plane lies in our line of sight, by observing the tiny dip in the starlight as the exoplanet transits across the star's disc, and so blocks a small fraction of its light.

Of these two techniques, careful use of the transit method offers the greatest hope of revealing an exomoon. If the exomoon passes in front of the star before or after the exoplanet, this will cause a tiny dip in the star's brightness just before or after the larger dip attributable to the planet itself. Furthermore, even if the dip in starlight caused by the exomoon's transit were not detectable, analysis of a series of repeated exoplanet transits might reveal fluctuations in their precise timing caused by the exoplanet's displacement to either side of the exoplanet–exomoon barycentre. An Io-like exomoon could be signalled by the presence of a plasma torus, whose passage across the face of the star could both dim its light and reveal itself through characteristic spectral lines. There might also be recognizable radio emissions that could be traced to such a plasma torus. Perhaps the strongest evidence for exomoons concerns a planet of an orange dwarf star known as J1407, revealed to be a 'super-Saturn' with a vast and spectacular ring system by multiple dips in the star's light as the rings transit across it. The gaps in the rings strongly suggest resonances with substantial moons, or the presence of less massive shepherd moons.

So far, none of these techniques has yielded definitive proof of exomoons. Colleagues working in the field of exoplanet research assure me that we may not be far off, though I expect that the first genuine detection of an exomoon may take some time to be confirmed beyond reasonable doubt.

Why do exomoons matter? Well, there are at least as many apparently habitable moons in our Solar System (Europa and Enceladus) as there are planets (Earth and Mars). If hydrothermal vents on ocean floors really are a good place for life to begin as suggested in Chapter 5, then icy exomoons with internal oceans throughout the galaxy could host life. As we have seen, the prospects for such life are reasonable even where the body's surface temperature is way below the freezing point of water.

Tidally heated icy exomoons could host life well beyond what has conventionally been regarded as a star's 'habitable zone', which requires liquid water at the surface. This may be mostly microbial life, and one might argue that it would be hard for any intelligent multicellular life to develop a technological civilization underwater, but it is feasible that there are many more inhabited moons than there are inhabited planets.

Internal oceans are not the only habitable environment that we can imagine for an exomoon. Exoplanet searches have revealed many giant exoplanets much closer to their star than Jupiter is to the Sun. Such a giant exoplanet could feasibly have an Earth-like exomoon, on which Earth-like life, and even intelligence, could arise. A famous example in recent science fiction is Pandora, the setting for the James Cameron movie *Avatar*, which orbits a giant exoplanet in its star's conventional habitable zone.

Maybe, if we ever do establish communication with aliens, they may be sceptical of our claim that we come from a world that goes directly round a star, rather than from one that orbits a giant planet.

Appendix: moons data

This table contains data for all the known moons of the planets and Pluto. Mean radius is quoted as a way to specify the size of a non-spherical moon by a single value.

Moons of the Earth and Mars

Name	Orbital radius/ 10^3 km	Orbital period/ days	Mean radius/ km	Mass/ 10^{20} kg	Density/ 10^3 kg m^{-3}
The Moon	384.4	29.53	1737	734.2	3.344
Phobos (Mars)	9.378	0.319	11.2	0.000106	1.90
Deimos (Mars)	23.46	1.26	6.1	0.000024	1.75

Jupiter's moons

Name	Orbital radius/ 10^3 km	Orbital period/ days	Mean radius/ km	Mass/ 10^{20} kg	Density/ 10^3 kg m^{-3}
Four inner	128–222	0.29–0.67	10–73	<0.075	<3.10
Io	421.6	1.769	1,822	893	3.50
Europa	670.9	3.551	1,561	480	3.01
Ganymede	1,070	14.97	2,631	1,482	1.94
Callisto	1,883	26.33	2,410	1,076	1.83
Fifty-nine outer	7,507–24,540	130–779	1–85	<0.095	

Saturn's moons

Name	Orbital radius/ 10^3 km	Orbital period/days	Mean radius/ km	Mass/ 10^{20} kg	Density/ 10^3 kg m^{-3}
Sixteen inner and co-orbital	134–377	0.56–2.74	16–93	<0.019	<1.300
Mimas	185.5	0.952	197	0.379	1.15
Enceladus	238.0	1.37	251	1.08	1.61
Tethys	294.7	1.89	528	6.18	0.985
Dione	277.4	2.74	561	11.0	1.48
Rhea	572.0	4.52	763	23.1	1.24
Titan	1,222	15.9	2,575	1,346	1.88
Hyperion	1,481	21.3	133	0.056	0.55
Iapetus	3,561	79.3	746	18.1	1.09
Thirty-eight outer	11,110–25,110	449–1,491	3–109	< 0.083	

Uranus' moons

Name	Orbital radius/ 10^3 km	Orbital period/days	Mean radius/ km	Mass/ 10^{20} kg	Density/ 10^3 kg m^{-3}
Thirteen inner	49.8–97.7	0.345–0.923	5–81		
Miranda	129.4	1.41	234	0.66	1.20
Ariel	191.0	2.52	578	13.5	1.67
Umbriel	266.3	4.14	528	585	1.40
Titania	435.9	8.71	561	789	1.71
Oberon	583.5	13.5	763	761	1.24
Nine outer	4,276–20,900	267–2,824	9–75		

Neptune's moons

Name	Orbital radius/ 10^3 km	Orbital period/days	Mean radius/ km	Mass/ 10^{20} kg	Density/ 10^3 kg m^{-3}
Six inner	48.2–105	0.294–0.950	18–102		
Proteus	117.6	1.12	208	0.5	
Triton	354.8	5.88	1,353	214	2.059
Nereid	5,513	360	170		
Five outer	15,730–48,390	1,880–9,374	20–30		

Pluto's moons (the sizes of some small moons are given as range estimates)

Name	Orbital radius/ 10^3 km	Orbital period/ days	Mean radius/ km	Mass/ 10^{20} kg	Density/ 10^3 kg m^{-3}
Charon	19.6	6.39	608	16.2	1.85
Styx	42	20.2	5–13		
Nix	48.7	24.9	21 × 18		
Kerberos	59.0	32.1	6–17		
Hydra	68.8	38.2	27 × 20		

Further reading

Books

If you want a book about planets at a similar level to this one, then I suggest:

D. A. Rothery, *Planets: A Very Short Introduction* (Oxford University Press, 2008).

There are many books about the Moon, and a few about some individual moons of other planets. I know of no recent book that attempts to describe moons in general in more detail than here. My own *Satellites of the Outer Planets: Worlds in Their Own Right*, 2nd edition (Oxford University Press, 2000) is now rather out of date but can still be found.

For the Moon in particular:

A. L. Chaikin, *A Man on the Moon: The Voyages of the Apollo Astronauts* (Penguin Books, 2007). At nearly 700 pages this is one of the most complete accounts of the Apollo programme.

A. Crotts, *The New Moon: Water, Exploration and Future Habitation* (Cambridge University Press, 2014). An excellent, recent account.

G. H. Heiken, D. T. Vaniman, and B. M. French (eds), *Lunar Sourcebook* (Cambridge University Press, 1991). A superb repository of lunar data and our understanding as it matured after Apollo. The complete book is viewable or downloadable free of charge at <http://www.lpi.usra.edu/publications/books/lunar_sourcebook/>.

S. Ross, *Moon* (Oxford University Press, 2009). A bulky 'coffee table' book, which has been favourably reviewed.

Books on other individual moons:

R. Greenberg, *Unmasking Europa* (Springer, 2008). A clear and authoritative account of Europa, with some scathing passages concerning the struggles to get the thin ice theory for Europa published in the face of establishment opposition.

R. Lorenz and J. Mitton, *Titan Unveiled: Saturn's Mysterious Moon Explored* (Princeton University Press, 2008). Published rather soon after Cassini's orbital tour of Saturn began, so check for a more recent specialist book on Titan.

M. Meltzer, *The Cassini–Huygens Visit to Saturn* (Springer, 2015). Mostly about mission planning and design, less than a quarter of the book is about the moons and rings.

J. Spencer and R. Lopes, *Io after Galileo: A New View of Jupiter's Volcanic Moon* (Springer, 2005). A thorough account published after the Galileo mission.

Online resources

There is a great deal of excellent and well-illustrated material about moons on the Internet, including galleries of images, short videos, and whole courses that you can study for free. The list below merely contains some of the highlights.

FutureLearn/Open University MOOC about moons (a free online course, three hours per week over eight weeks) <https://www.futurelearn.com/courses/moons>.

Virtually the same free course on an Open University site, to study at your own pace, but with less support <http://www.open.edu/openlearn/science-maths-technology/moons/content-section-overview>.

A 'Moon trumps' card game, to play online and test your knowledge of moons against a computer <http://www.open.edu/openlearn/moontrumps>.

Four Open University videos about moons <http://www.open.edu/openlearn/science-maths-technology/science/across-the-sciences/moons-the-solar-system>.

Moon myths debunked in a short animated video <http://www.open.edu/openlearn/moonmyths>.

Another debunking of supermoons and a demonstration of the 'Moon illusion' <http://www.universetoday.com/118990/why-does-the-moon-look-so-big-tonight/>.

A video recording of a fifteen-minute lecture I gave about life in moons for Word Space Week in 2014 <https://www.youtube.com/watch?v=95RpUMHqw-0>.

A gallery of NASA images of all planets and their moons <http://photojournal.jpl.nasa.gov/>.

'Quasi-moon' horseshoe orbits and the orbit of Cruithne can be examined at <http://www.astro.uwo.ca/~wiegert/3753/3753.html>.

A radar video of the smallest known moon, orbiting asteroid 2004 BL$_{36}$ <http://www.jpl.nasa.gov/news/news.php?feature=4459>.

High-resolution views of the Apollo landing sites, showing footprints and rover tracks from the Lunar Reconnaissance Orbiter <http://www.lroc.asu.edu/featured_sites/>. Use the 'Flip Book' option to see how shadows change during the lunar day. You can also see other landers and rovers, and some new craters.

See how the Moon must rotate once per orbit to keep the same face permanently towards the Earth at <http://www.open.edu/openlearn/RotatingMoon>.

Animations showing the Moon's phases, libration, and changing apparent size over the course of a year <http://svs.gsfc.nasa.gov/cgi-bin/details.cgi?aid=4236>.

You can find a movie of an eruption of Tvashtar on Io if you visit <http://pluto.jhuapl.edu/Multimedia/Videos/index.php> and search under 'Data Movies' and then 'New Horizons at Jupiter'.

A three minute video about eruptions on Io and Enceladus <https://www.youtube.com/watch?v=ZVybNKuhpSY>.

Images, and other products, from the Cassini mission exploring Saturn and its moons <http://saturn.jpl.nasa.gov/>.

Waltz Around Saturn, a movie made from successive Cassini images and brilliantly set to music <http://vimeo.com/70532693>.

A movie of Prometheus and Pandora shepherding Saturn's F ring <http://saturn.jpl.nasa.gov/video/videodetails/?videoID=95>.

Voyager Uranus flyby movie <https://www.youtube.com/watch?v=DrKQaDupdWQ>.

Voyager Triton flyby (based on processing in 2014) <http://www.lpi.usra.edu/icy_moons/neptune/triton/movie/index.shtml>.

Phobos passing across Deimos, as seen from the surface of Mars <http://photojournal.jpl.nasa.gov/catalog/PIA17089>.

The New Horizons website, showing fly-by images of Pluto's moons <http://pluto.jhuapl.edu/>.

An audacious proposal to scatter 'chipsats' onto Europa is described here, though without mentioning any planetary protection considerations <http://www.astrobio.net/news-exclusive/swarm-tiny-spacecraft-explore-europas-surface-rapid-response>.

“牛津通识读本”已出书目

古典哲学的趣味
人生的意义
文学理论入门
大众经济学
历史之源
设计，无处不在
生活中的心理学
政治的历史与边界
哲学的思与惑
资本主义
美国总统制
海德格尔
我们时代的伦理学
卡夫卡是谁
考古学的过去与未来
天文学简史
社会学的意识
康德
尼采
亚里士多德的世界
西方艺术新论
全球化面面观
简明逻辑学
法哲学：价值与事实
政治哲学与幸福根基
选择理论
后殖民主义与世界格局

福柯
缤纷的语言学
达达和超现实主义
佛学概论
维特根斯坦与哲学
科学哲学
印度哲学祛魅
克尔凯郭尔
科学革命
广告
数学
叔本华
笛卡尔
基督教神学
犹太人与犹太教
现代日本
罗兰·巴特
马基雅维里
全球经济史
进化
性存在
量子理论
牛顿新传
国际移民
哈贝马斯
医学伦理
黑格尔

地球
记忆
法律
中国文学
托克维尔
休谟
分子
法国大革命
民族主义
科幻作品
罗素
美国政党与选举
美国最高法院
纪录片
大萧条与罗斯福新政
领导力
无神论
罗马共和国
美国国会
民主
英格兰文学
现代主义
网络
自闭症
德里达
浪漫主义
批判理论

德国文学
戏剧
腐败
医事法
癌症
植物
法语文学
微观经济学
湖泊
拜占庭
司法心理学
发展
农业
特洛伊战争
巴比伦尼亚
河流
战争与技术
品牌学
数学简史
物理学

儿童心理学
时装
现代拉丁美洲文学
卢梭
隐私
电影音乐
抑郁症
传染病
希腊化时代
知识
环境伦理学
美国革命
元素周期表
人口学
社会心理学
动物
项目管理
美学
管理学
卫星

电影
俄罗斯文学
古典文学
大数据
洛克
幸福
免疫系统
银行学
景观设计学
神圣罗马帝国
大流行病
亚历山大大帝
气候
第二次世界大战
中世纪
工业革命
传记
公共管理
社会语言学